Biology Unmoored

Biology Unmoored

MELANESIAN REFLECTIONS ON LIFE AND BIOTECHNOLOGY

Sandra Bamford

UNIVERSITY OF CALIFORNIA PRESS

BERKELEY   LOS ANGELES   LONDON

University of California Press, one of the most distinguished university presses in the United States, enriches
lives around the world by advancing scholarship in the humanities, social sciences, and natural sciences. Its activities
are supported by the UC Press Foundation and by philan-
thropic contributions from individuals and institutions.
For more information, visit www.ucpress.edu.

University of California Press
Berkeley and Los Angeles, California

University of California Press, Ltd.
London, England

Library of Congress Cataloging-in-Publication Data

Bamford, Sandra C., 1962–
    Biology unmoored : Melanesian reflections on life and
biotechnology / Sandra Bamford.
        p.   cm.
    Includes bibliographical references and index.
    ISBN: 978-0-520-24712-3 (cloth : alk. paper)
    ISBN: 978-0-520-24713-0 (pbk. : alk. paper)
        1. Hamtai (Papua New Guinean people)—Ethnobiol-
ogy. 2. Hamtai (Papua New Guinean people)—Agriculture. 3. Hamtai (Papua New Guinean people)—Psychol-
ogy. 4. Body, Human—Social aspects—Papua New
Guinea—Gulf Province. 5. Indigenous peoples—
Ecology—Papua New Guinea—Gulf Province.
6. Ethnobiology—Papua New Guinea—Gulf Province.
7. Biotechnology. 8. Genetic engineering. 9. Gulf Province
(Papua New Guinea)—Social life and customs.   I. Title.

    DU740.42.B34   2007
    305.89'912—dc22   2006009333

Manufactured in the United States of America

16   15   14   13   12   11   10   09   08   07
10   9   8   7   6   5   4   3   2   1

*To my father and mother. Whatever kinship may be,
I could not have asked for better parents.*

CONTENTS

ILLUSTRATIONS

FIGURES

MAPS

ACKNOWLEDGMENTS

My doctoral research was made possible by fellowships from the Social Sciences and Humanities Research Council of Canada, as well as a research grant from the Wenner-Gren Foundation for Anthropological Research. I am deeply grateful to both of these organizations for their financial support of my fieldwork in Papua New Guinea. I would also like to thank the Graduate School of Arts and Sciences at the University of Virginia for its support while I was writing an initial draft of this work in the form of a Dean's Alumni Dissertation Year Fellowship. Subsequent funding by the University of Toronto allowed me to rethink substantial portions of my Melanesian fieldwork in light of recent developments in kinship studies and a growing literature in science and technology.

I am deeply indebted to many people for their intellectual and emotional support over the years. For instilling in me a love of anthropology, and for his ongoing support and encouragement for over the years, I would like to thank Michael Lambek. He has been a true friend and mentor. I can only hope to achieve a fraction of the positive impact on my own students in the years to come that he has had upon the course my own life has taken. The influence that Roy Wagner has had on my thinking has been nothing short of profound; it will be evident throughout this work, not to mention how I will think about the discipline of anthropology in the years to come. I am grateful to him for his patient guidance, support, and encouragement while he acted as my dissertation supervisor at the University of Virginia, and for providing me with an exemplary model of what it means to do ethnographic research. To Susan McKinnon, I owe an enormous intellectual debt for serving as a source of infallible advice and critical commentary over the years. In addition to always pushing me both theoretically and analytically, she has kept me sane at times when I was sure that sanity was next to impossible. I would also like to acknowledge a huge debt to Frederick Damon for keeping me intellectually honest, and for encouraging me to ask (even if I cannot always answer) the tough questions. Margaret Holmes Williamson kindly read this work when it was in its very initial stages, and provided me with many valuable insights that I have subsequently tried to incorporate. The influence that Marilyn Strathern has had on my thinking will be apparent throughout this work. The intellectual debt that I owe to her is obvious. What is perhaps not so obvious are the many kindnesses that she has shown me over the years, and for no apparent reason except a true generosity of spirit. By extending a warm and welcoming hand to a junior scholar, she has provided me with an exemplary model of what true intellectual kindness means. I also extend a special thanks to my fellow Angan researchers: Pascale Bonnemère, Maurice Godelier, Gilbert Herdt, Pierre Lemonnier, and Jadran Mimica, all of whom have helped my research in innumerable ways. In addition, I wish to thank all of the anonymous reviewers who read this work when it was submitted for publication consideration; they helped immeasurably in making this account far more readable than it would have been otherwise.

Abigail Adams, David Akin, Rafael Alvarado, Steven Gaetz, Teresa Holmes, Meena Khandelwal, Bruce Koplin, Lynn Koplin, Ken Little, Peter Metcalf, Catherine Moffatt, Rebecca Popenoe, Cindy Robins, Joel Robbins, Margo Smith, Karen Sykes, Daphne Winland, and Laurie Zadnik have helped in more ways than I can possibly count. I am indebted to each of them for their support, friendship, humor, and the depth of their analytic insights. Catherine Moffatt read this manuscript from beginning to end in various drafts (and in some sections, many times). Her analytical insights and keen editorial eye have helped immeasurably in making the final product more coherent than it otherwise would have been. I am also hugely grateful to Sikata Banerjee, Lori Beaman, Norman Buchinani, Audrey Glasbergen, Doreen Indra, Jennifer Johnson, Catherine Kingfisher, Paul Letkemann, Rakesh K. Ratti, Renee Soloudre La-France, and Donna Young, whose support, help, and encouragement provided me with the sustenance I needed to complete this work.

Without the unfailing support of my parents—Anne and Robert Bamford—and my sisters—Leslie, Heather, and Barb—I would have lacked the courage to begin this project and the stamina to finish it. Words cannot express the debt of gratitude that I owe them. David Main has proved to be an unerring source of intellectual, emotional, and comedic support over the years (not to mention serving as a superlative computer consultant). Mathew and Christopher remind me of what it means to be human. Rob Rainford came into my life as this project was wrapping up. His love and support nurtured my spirit through the final stages of completing this work. And I must extend my heartfelt thanks to Tess, the Labrador retriever I adopted in Papua New Guinea almost fifteen years ago, and who continues to warm my feet as I type these words.

I wish to thank Hans and Beatte Grauvogl, who took me in when I first arrived in Papua New Guinea. I am also indebted to Norman Sorari, the district officer at Kaintiba, and to Father John Flynn of the Catholic Mission Station at Hawabango. Boni Laclucis provided me with friendship and a place to stay whenever I was in Kerema. Tim helped to make a wonderful stay in PNG even more memorable. The provincial government of Gulf Province was consistently supportive of my research, and to them I extend my heartfelt thanks. Finally, it is to Kamea who I owe my biggest debt. For the warmth, hospitality, and kindness that they have shown me over the years, my gratitude and appreciation knows no bounds.

Parts of chapter 1 appeared in slightly different form in "Humanized Landscapes, Embodied Worlds: Land and the Construction of Intergenerational Continuity among the Kamea of Papua New Guinea," *Social Analysis* 42, no. 3: 28–54, copyright © 1998 Berghahn Books; and parts of chapter 5 appeared previously in "On Being 'Natural' in the Rainforest Marketplace: Science, Capitalism, and the Commodification of Biodiversity," *Social Analysis* 46, no. 1 (2002): 35–50, copyright © 2002 Berghahn Books. Both reprinted with the permission of Berghahn Books.

Parts of chapter 2 appeared in different form in "Conceiving Relatedness: Non-Substantial Relations among the Kamea of Papua New Guinea," *Journal of the Royal Anthropological Institute* 10, no. 2 (2004): 287–306. Reprinted with the permission of Blackwell Publishing.

Parts of chapter 3 originally appeared in different form in "To Eat for Another: Taboo and the Elicitation of Bodily Form among the Kamea of Papua New Guinea," in *Bodies and Persons: Comparative Perspectives from Africa and Melanesia,* edited by Michael Lambek and Andrew Strathern,

Introduction
Conceptual Frameworks

Once the genies let the babies into the bottle, it may be impossible to get them out again.

LEON KASS

(1989:347)

TEST-TUBE BEGINNINGS

September 12, 1973. Columbia Presbyterian Medical Center, New York

At eleven o'clock in the morning, Landrum Brewer Shettles stood in the foyer of this esteemed institution waiting for someone to deliver a package. A recipient of the prestigious Markle Scholarship, Shettles was an established, if somewhat eccentric, New York fertility doctor known for publicizing a number of low-tech methods that he claimed couples could use to predetermine the sex of their baby. The so-called Shettles method entailed everything from taking hot baths to wearing loose underwear to subjecting the prospective mother to unlikely douches with baking soda (to yield a boy) or vinegar (to yield a girl) (Mundy 2004). On this particular day, Shettles had loftier things on his mind. In addition to his work on sex selection, Shettles was trying to gain even greater control over the process of reproduction. In particular, he was experimenting to see whether it might be possible to unite egg and sperm outside the womb. The package he was waiting for could make or break his career.

A few days earlier, Doris Del-Zio and her husband, John, checked in at New York Hospital, luggage in hand. It was a well-worn routine for the Florida couple. On three previous occasions, Doris had visited her fertility specialist, William Sweeney, who had tried to remove obstructions from

I

Doris's blocked fallopian tubes (Henig 2004: 1). None of these attempts had succeeded. The doctor had also made several attempts at artificial insemination using John's sperm. Again, these efforts failed to result in a baby (Henig 2004; Weinberg 2004).

On this particular occasion, the couple was hoping to make medical history. Working with Sweeney and Shettles, the Del-Zios had agreed to try an experimental procedure that until then had been tried only on mice and rabbits. A few of Doris's eggs would be surgically removed and fertilized in a glass test tube with John's sperm. If one of the sperm succeeded in fertilizing one of the eggs, the resulting zygote would be placed in an incubator and allowed to grow. It would take approximately three days for the fertilized egg to develop into the thirty-two-cell ball known as a morula, and another day or two to grow into a blastocyst—a fluid-filled sphere composed of a few hundred cells (Henig 2004: 2–4). At this point the embryo would be ready to attach itself to the uterine wall and could be surgically transferred to Doris's uterus. The proposed undertaking was tricky and would be rendered even more difficult by geography (Henig 2004: 3). Sweeney knew how to harvest eggs. Shettles had the expertise to fertilize them in vitro. Separating these two men were one hundred city blocks, and if the plan was to succeed, time was of the essence.

At approximately five past eleven on that September morning, John Del-Zio arrived, breathless, in the lobby of Presbyterian Hospital. Stashed in his coat pocket, encased in protective bubble wrap, were two sterile test tubes that contained his wife's eggs along with some of her ovarian secretions (Andrews 1999b: 16). He handed the tubes to Shettles, who carefully transported them to his laboratory on the sixteenth floor. There, the physician got to work. He pulled out a new, sterile test tube in which he mixed Doris Del-Zio's eggs, half a cubic centimeter of sterile human placental blood for nourishment, and several drops of John Del-Zio's freshly collected sperm. As he was to describe it later, the contents looked a bit "murky," but Shettles felt certain that the test tube contained at least one mature, fertilizable egg and at least one motile sperm that was ready to do the fertilizing (Henig 2004: 23). He sealed the test tube with a black rubber stopper, balanced it in a beaker, and placed it in an incubator set at ninety-eight degrees Fahrenheit. Happy and confident, content with his day's work, Shettles stepped out in the hallway, eager to tell his colleagues about the test-tube baby that was developing in his laboratory (Henig 2004: 23).

The experiment could have worked. In the opinion of Sweeney and Shettles, it *should* have worked. Had things turned out as planned, Doris Del-

Zio would have given birth to the first human baby that had been created through in vitro fertilization (IVF), and Lesley Brown—the woman who currently holds that distinction—would have lived a life out of the media spotlight.[1] But it was not to be.

Early the next morning Shettles responded to a page on his beeper. He was ordered to report to the office of his supervisor, Raymond Vande Wiele, chair of the Department of Obstetrics and Gynecology at Columbia Presbyterian Hospital and Medical Center (Andrews 1999b: 16). Through the departmental grapevine, Vande Wiele had heard about the procedure that was taking place on the sixteenth floor, and he was none too pleased. He was furious that Shettles had attempted this kind of experiment without first seeking the hospital's approval, arguing that it was unethical and immoral to carry on this work. "How far can we go?" he demanded (Maeder 2001). He also claimed that he was worried that the contents of the test tube had been contaminated and that he did not want to see Doris Del-Zio risk infection, possibly even death, if the contents were transferred to her womb (Andrews 1999b: 16). As the two men exchanged words, Shettles noticed the Del-Zio test tube sitting on a coffee table in Vande Wiele's office. The department chair had removed it from the incubator. The stopper had been taken out, and the mixture had been kept at room temperature for several hours, thereby destroying the experiment and dashing the couple's hopes of giving birth to a child of their own (Andrews 1999b: 16–17). Within days, an executive committee at the hospital met and voted unanimously to ask for Shettles's resignation (Henig 2004: 103). Doris Del-Zio was still in the hospital recovering from egg-removal surgery. When the news of Vande Wiele's action reached her, she sank into a deep depression (Andrews 1999b: 17). One year later, she and her husband filed suit against Columbia Presbyterian Medical Center. The case was decided in August 1978: the Del-Zios were awarded fifty thousand dollars in damages (Andrews 1999b: 17; Wallis 1984).[2]

What is noteworthy about the Del-Zio case from the perspective of the twenty-first century is the horror with which IVF was viewed when it was first proposed. In 1973, the year of Sweeney and Shettles's failed experiment, many people felt that IVF posed a serious threat to the very fabric of human existence: marriage, fidelity, family relations; our sense of who we are and where we came from; what it means to be human, normal, acceptable; ideas about love, sex, gender, and nurturance; our conception of our place within the organic world—all these and more would be threatened if IVF were allowed to proceed (Henig 2004: 6). A Harris poll published in 1969 had found that more than half of North American adults believed that the emerging reproductive technologies (i.e., in vitro fertilization, artificial insemination, and surrogate motherhood) were "against God's will" and would encourage promiscuity (Rosenfeld and Harris 1969: 54). More than two-thirds of respondents believed these techniques would "emasculate men" and signify "the end of babies born through love." Public sentiment was the same in Great Britain, where many people thought that fertilizing a human egg in the laboratory would be crossing a critical line. The British magazine *Nova* ran a cover story in the spring of 1972 suggesting that test-tube babies were "the biggest threat since the atom bomb." The authors asked the public to take a stern stance and regulate the "rogue" scientists (Henig 2003b: 64). Even the Nobel Prize–winning molecular biologist James Watson warned a U.S. congressional subcommittee that "all hell [would] break loose" if embryo transplantation were allowed to move forward. "The nature of the bond between parents and the bond between parents and children . . . and everyone's values about his individual uniqueness," he declared, "could be changed beyond recognition." He urged Congress to establish a commission to consider the ramifications of test-tube conceptions and embryo transplants and possibly even to "take steps quickly" to make them illegal (quoted in Rorvik 1971).[3]

Since scientists first learned how to fertilize human eggs in labs, innovations in biotechnology have continued to flourish. IVF has become routine, now practiced as an outpatient procedure at more than four hundred clinics across the United States alone. Its reach has also broadened beyond all early predictions, making biological fathers out of dead men and allowing women to bear children well past menopause (Harris 1969; Mestel 2003). In addition to advances in the field of reproductive medicine, the past few decades have witnessed the advent of several innovations that would have been unthinkable earlier. These include cloning, genetic engineering, transgenic species crossings, and the potential to culture human embryos as sources of "replacement parts." Indeed, recent applications of biotechnology have been so profound and sweeping that the *Washington Post* has declared "the age of reproductive biology" to be the successor to the nuclear age, "the next scientific advance to shake the foundations of our social values" (quoted in Henig 2004: 211).

Yet, if the world today seems substantially different from the one that existed when Sweeney and Shettles tried their experiment, one thing remains the same. The societal fears that preceded the birth of the first test-tube baby are nearly identical to the fears that exist today concerning the use of other reproductive technologies (Henig 2003a: A27; Mestel 2003; Pence 1998). Consider the views of Leon Kass, a bioethics professor at the University of Chicago, who has been weighing in on debates concerning the uses of science and technology for the last twenty-five years (Henig 2003). In 1979, only months after the birth of the first test-tube baby in the United Kingdom, Kass stated before a meeting of the U.S. government's Ethics Advisory Board: "More is at stake [with IVF research] than in ordinary biomedical research or experimenting with human subjects at risk of bodily harm. By tampering with and confounding these origins, we are involved in nothing less than creating a new conception of what it means to be human" (quoted in Henig 2003a: 27). Kass heads President George W. Bush's Ethics Advisory Committee (see chapter 4), and when he talks about recent applications of biotechnology, he uses many of the same words and phrases that he used in opposing the idea of test-tube babies during the 1970s. Kass, like many others, sees calamity and moral ruin as the inevitable outcomes of creating life through nonsexual means (Henig 2003a: 27).[4]

Biology has long served as an orienting device for Europeans and North Americans. Since Darwin first published *The Origin of Species* in 1859, a biological paradigm has furnished Western audiences with a set of tropes through which we have understood our relationship to other human beings and to nonhuman species. More than simply defining Euro-American kinship configurations, a biological framework has figured centrally in how Westerners define bodies, persons, gender, ethnicity, and the place of human beings in the organic world. One of the reasons that recent innovations in science and technology have proven to be so disturbing for so many is that they prompt us to imagine a world in which biology no longer performs its traditional grounding functions. Paul Rabinow (1996) coined the term *bio-sociality* to describe a transformation in which the biological— and, in particular, the new genetics—are no longer modeled on nature, but have come to be modeled on culture and thus are understood to be "artificial" (cf. Franklin 2001b: 303). Similarly, Marilyn Strathern (1992a, 1992b) documents the "vanishing" of taken-for-granted assumptions about natural processes and the emergence of a domain in which "nature" is increasingly subject to consumer choice (cf. Franklin 2001b). For many Europeans and North Americans, biology no longer defines a set of time-honored and immutable distinctions. What we are confronted with, instead, is a world in which biology has become unmoored.

This book is a detailed analysis of what it means to live in a world that is not structured in terms of biological thinking. Between August 1989 and

February 1992, I conducted ethnographic research with Kamea, a highland people who occupy the rugged interior of the Gulf Province, Papua New Guinea. In contrast to Europeans and North Americans, Kamea do not rely upon physiological reproduction as a means of grounding social identities and relationships. Although they are quite articulate when it comes to specifying their views about what goes into the making of a baby—a father's semen must mix with a mother's blood—this is not used as a means of tracking social relationships through time. Despite my repeated attempts to anchor intergenerational relations in a procreative bond, Kamea were quite insistent that parents do not share any kind of physical connection with their offspring. Kamea do have a means of tracing social relationships through time, but this is not seen to rest upon genealogical connections; instead, it eventuates from the ties that people form with the land.

As we shall see, this system of ideas structures more than local kinship configurations. To the extent that this framework engenders a radically different view of life, it comes to be manifest in a diverse array of contexts. During the thirty-one months that I lived and worked among Kamea, I became acquainted with a world in which the following tenets were taken as axiomatic:

- The capacity to mother or father a child is not given in the nature of things, but depends, instead, on the relationships that one forms with nonhuman species.

- In contrast to a guiding precept of evolutionary biology, there is no embodied link that connects the generations; the organic world is understood in nongenealogical terms.

- Males and females, mothers and fathers, create markedly different kinds of social relations. Unlike Euro-American logic, there is no essential equivalence with respect to the *kinds* of relationships that men and women are seen to engender.

- Bodies do not exist as autonomous entities, but have the capacity to act directly upon one another. Therefore, it is entirely possible for one person to eat for another.

In this book, I unpack the logic of these ideas. More specifically, I use Kamea conceptions as a counterreflexive voice through which to consider many unexamined tenets of a biological framework, including how it constructs the relationships between male and female, "nature" and "culture," bodies and persons, primordial origins and the passage of time. Insomuch as these concerns have figured centrally in the development of anthropological theory, this book also explores the extent to which the "social" and "natural" sciences have constituted themselves through a set of metaphorical borrowings. In addressing these themes, I give serious attention to the moral and political implications of extending biology as a worldview to other parts of the globe—an occurrence that is increasingly brought into practice through such initiatives as the Human Genome Diversity Project and the growing call for global biodiversity conservation.

ANTHROPOLOGICAL CONCEPTIONS AND MISCONCEPTIONS

To the extent that this work critically interrogates the link between kinship and physiological reproduction, it engages with an important series of debates in the anthropological literature. Since its inception as an academic discipline well over a century ago, the study of kinship has been central to the anthropological tradition. Nineteenth-century accounts of comparative social organization focused much of their attention on documenting divergent beliefs about procreation, or what Euro-Americans informally called "the facts of life" (Franklin and Ragoné 1998: 1). For many early anthropologists, acquiring "accurate knowledge" of how offspring were produced signaled a crucial stage in the transition from "savagery" to "civilization," characterized by the triumph of "reason" over "nature" (Franklin 1998a: 102). The publication in 1916 of W. H. R. Rivers's seminal essay, "The Genealogical Method of Anthropological Inquiry," further solidified the attention that was given to reproductive knowledge as the basis for cross-cultural comparison. As Mary Bouquet has noted, Rivers intended nothing less than to establish ethnography as a science "as exact as physics or chemistry" (Bouquet 1993: 114). Toward this end, Rivers enjoined fieldworkers to obtain "basic information on relatedness" by collecting genealogical data as a standard component of ethnographic research.[5] In the wake of his publication, generations of anthropologists set off for distant parts of the globe, armed with a methodological tool kit with which they thought they could build a truly comparative science of man.

It was, of course, David Schneider who came to question the foundation upon which over half a century of anthropological research had been based. In his first major work, *American Kinship: A Cultural Account* (1968), Schneider set out to expose the system of ideas that structured North American kinship configurations. He argued that North American kinship rested upon a distinction between two orders: that of nature and that of law. For North Americans, one can be related "by blood" (in "nature"), or one can be related "in law" (by marriage). Sexual intercourse served as a core symbol in this conceptual framework insomuch as it provided a bridge between these two domains:

> The family is defined by American culture as a "natural" unit which is "based on the facts of nature. . . ." The fact of nature which the cultural construct of the family is based is . . . that of sexual intercourse. This figure provides all of the cultural symbols of American kinship. The figure is formulated in American culture as a biological entity and a natural act. Yet throughout, each element which is culturally defined as a natural is at the same time augmented and elaborated, built upon and informed by the rule of human reason, embodied in law and in morality. (Schneider 1968: 33–40)

Sexual intercourse between two people who are biologically unrelated (but, legally—i.e., socially—married) results in children who are connected to their parents through shared biogenetic substance (nature), but who will relate to one another on the basis of love (i.e., social conventions). Kinship, in other words, is biology with culture put on top. It has to do with the social regulation of biological givens.

To the extent that Schneider demonstrated that our taken-for-granted understandings of kinship were culturally constructed, the implications of his study were far-reaching. Effectively, he had suggested that North Americans drew upon a distinction between "social" and "biological" kinship to classify different kinspersons in their social universe. Even more importantly, he suggested that because these categories had informed ethnographic research abroad, we had failed as a "science" to understand much of anything about the world. For several decades, cultural anthropology had staked out a unique place in the social sciences by virtue of its seemingly obsessive interest in matters pertaining to kinship and succession (Franklin 1998a: 102). Any particular kinship system was seen to be a cultural elaboration on the basic "facts" of human reproduction. It was the task of sociocultural anthropology to document the different interpretative spins that each society placed on procreative arrangements. Yet, if the distinction between "social" versus "biological" kinship was a culturally specific one, then what exactly was it that ethnographers had been investigating?

In his second major work, *A Critique of the Study of Kinship,* Schneider (1984) tackled this question in detail. He argued that what anthropologists had been studying (or thought they had been studying) for well over a century was nothing more than a reflection of our own cultural baggage taken abroad. The assumption that people everywhere assigned cultural significance to the "facts" of human heterosexual reproduction, and classified relatives in accordance with this principle, revealed more about the internal workings of Euro-American cultural logic than it did about the societies that anthropologists purported to be investigating.

In the course of deconstructing the universal basis of kinship, Schneider simultaneously deconstructed kinship as an independently existing analytical domain (Weston 1995: 89). After the publication of *Critique,* it was no longer possible to assume that the study of kinship was central to anthropology. Having "dismantled the subfield's procreative underpinnings" (Weston 1995: 89), Schneider had undermined the foundations upon which kinship as a domain of theorizing rested. He did so by demonstrating that kinship was not the "same thing" in all cultures. If this was the case, then the comparative mission of anthropology had failed (Carsten 2000: 25), since like was not being compared with like. Having lost its genealogical underpinnings, the study of kinship appeared to lose its status as an independent object of analytical inquiry.

In the wake of Schneider's devastating critique, anthropologists were left with something of a conundrum: they could either abandon the concept of kinship altogether, or they could adopt a far wider definition of the concept than had been used in previous discussions. Over the past decade, an attempt has been made in the latter direction. This effort has drawn upon a notion of "relatedness" (see, for example, Carsten 2000) to define kinship as a "process" rather than a state of being. Proponents of this model have given much attention to optative and adoptive relations, to postnatal modes of creating substance-based links through purposeful acts of feeding, caring, loving, and sharing (see Viveiros de Castro [forthcoming] for a discussion of this position). Here, the overwhelming emphasis is on the "socially created" nature of consanguineal relations (Viveiros de Castro forthcoming), on the ability to create bonds of corporeal consubstantiality through conscious human effort. Ladislav Holy champions the utility of this framework in a recent textbook intended for introductory students:

To be meaningful as a concept, kinship has to be understood as a culturally specific notion of relatedness *deriving from shared bodily and/or spiritual substance and its transmission.* In the West as well as in many non-Western cultures, this may be seen as resulting from the process of sexual reproduction. In yet other cultures, this may well be seen as resulting from sharing the same food, living on the same land, or whatever. In Western terms, the latter processes would of course not be classified as biological but social and therefore not "kinship" within the traditional anthropological definition . . . If the separation of the "social" from the "biological" which underlies the traditional anthropological definition of kinship is a specifically Western contrast which is not drawn in all cultures, the adherence to the conceptualisation of kinship as shared substance and its transmission . . . would have more far-reaching consequences for our analytical practice than we have so far realised. (Holy 1996: 168–71; italics added)

This model has been set forth in an attempt to avoid the well-worn dichotomy between "social" versus "biological" kinship. The point is to show that the boundaries between the social and the biological are more permeable than Western commonsense assumptions would have us believe. Yet, in managing to escape the constraining influence of one set of Euro-American assumptions, others remain tellingly intact. Perhaps most significantly, kinship continues to be seen as a material relationship that is expressed in corporeal terms.

In the chapters that follow, I challenge the hold that a procreative (or, pseudoprocreative) model has on Euro-American understandings of relatedness. More specifically, I provide the first detailed account of a world in which consanguineal relations are not imagined to entail an embodied connection, whether created at birth or through postnatal means.[6] In contrast to the guiding precepts of biology—a framework that implicitly underpins a great deal of modern anthropological theory—Kamea posit a marked distinction between what goes into the *making* of persons in a physical sense and what *connects* them through time as social beings. Here, intergenerational relations are not based on any type of bodily connection. This carries with it a number of important implications. Not only are birth and parentage radically different when seen through Kamea eyes, so too is the link between human beings and the nonhuman environment.

Throughout this book, I use Kamea ideas to address three interrelated themes. First, I am concerned with highlighting the extent to which a great deal of theory in the social sciences implicitly takes a biological paradigm as its underlying foundation. In the anthropological literature, for example, matrilineal payments have frequently been interpreted as an attempt to "buy off" the reproductive claims of one parent, so that the product of the sexual union—the child—can be recruited to the descent group of the other (M. Strathern 1987; Wagner n.d., 1967, 1977; J. Weiner 1982). Our theory of mortuary rites rests upon a similar set of suppositions. Death rituals are commonly said to be about "deconceiving" an individual so that the procreative substances a person received at birth can be sent back to the places from which they came (Mosko 1983; Munn 1986; A. Weiner 1980). However illuminating these interpretations have been in many cases, they cease to have explanatory power in a world where procreation is not seen to be the underlying basis of social connections. Here, it becomes necessary to seek alternative interpretative venues for a wide range of practices—a task that is undertaken in this book with respect to Kamea.

Second, this book illustrates the need to bring together two fields of inquiry that are of growing importance to Western academics and policy makers: studies of kinship and the investigation of human-environmental relations. Although treated as separate fields in the past, this book demonstrates the utility of considering these areas in relation to one another, by virtue of their mutual embeddedness within a biological framework. In the folk model of the West, physiological reproduction not only grounds human "kinship," but it serves as the defining criterion of species membership. Biological forms that can mate with one another and produce reproductively viable offspring are said to be of the same species (Mayr 1982). This system of ideas has several important implications. Perhaps most significantly, it precipitates an unbreachable gap between human beings and all other life forms on the planet. To the extent that supposedly unique reproductive histories are used to demarcate so-called species boundaries, a biological framework leaves human beings estranged from other constituents of the organic world. The world seen through biological eyes is populated by various categories of beings, each hermetically sealed from all others in a constitutive sense. In this book, I reveal the contours of an alternative conceptual framework. For Kamea, human reproductive potential is *directly tied* to other species, rather than defined in contradistinction to them. It is only by forming intimate relations with nonhuman constituents of the organic world that human kin-based relationships are created.

Finally, this work takes seriously the implications of extending biology as a worldview to other parts of the globe. Recent innovations in the field of science and technology have not been confined to Western countries; increasingly, these developments are becoming part of the global flow that crosses national and geopolitical boundaries.[7] The commodification of genetic resources, the mapping of the human genome, and the manufacture and sale of genetically modified organisms (GMOs) by northern countries to southern ones all represent important venues through which a biological paradigm is being exported to non-Western audiences. In this book, I consider the social and political implications of what might be called an emerging form of biological imperialism. Using Kamea conceptions as an analytical springboard, I suggest that the global traffic in genes will have far-reaching consequences above and beyond its anticipated commercial effects, holding that it will fundamentally alter how many indigenous and Fourth World peoples conceptualize self, sociality, and processes of life more generally.

OUTLINE OF THE BOOK

Throughout this book, I adopt a mode of exposition that tacks back and forth between Kamea ethnography and recent developments in the fields of science and technology. Each chapter opens with a consideration of some recent development that has been making headline news in Europe and North America, such as cloning, assisted conception technologies (ACTs), the mapping of the human genome, transgenic species crossings, and the emergence of a discourse based on fetal rights. Each of these developments is selected because it helps to reveal a particular side-effect of biological thinking; namely, how it constructs species, time, kinship, bodies, persons, gender, and ethnicity. These innovations are then placed in deliberate juxtaposition with a particular aspect of Kamea ethnography. My goal in adopting this technique is to use Kamea conceptions as a destabilizing device through which to view recent shifts in Western kinship configurations, including how Euro-Americans perceive their relationship to other life forms. The end result of this technique is the creation of a three-way dialogue in which the assumptions of the "new biology" challenge those of the "old biology," and Melanesian perspectives challenge those of the West.

In addressing these themes, my argument unfolds as follows. In chapter 1, I turn my attention to a founding problematic of evolutionary biology: the relationship between human beings and other constituents of the organic world. I document the extent to which a culturally specific model of biology underpins Euro-American understandings of human-environmental relations. This chapter opens with a consideration of recent controversies in the West about the use of genetically modified organisms. Through an examination of the rhetoric that has structured both sides of this debate, I suggest that much of the negative reaction that has accompanied the use of these forms stems from a perception on the part of Europeans and North Americans that crossing so-called species boundaries is not only "wrong," but potentially dangerous as well. In the second half of this chapter, this system of ideas is juxtaposed with Kamea conceptions of the world. Here, I show that the elicitation of human social life is *directly tied* to the reproductive cycles of other creatures, rather than being defined in contradistinction to them. Through the planting of select trees, intergenerational links are formed between men who work the same ground, not on the basis of common descent, the underlying precept of a genealogical framework, but rather through their ability to share in the creation of a humanized landscape. People do not hold land so much as it comes to hold them, thus eliciting a sense of generational continuity.

If chapter 1 focuses on the relationship between human beings and other constituents of the organic world, chapter 2 addresses the links among human beings themselves. Here, I explore Western ideas of heterosexual reproduction, including the assumption that it results in an embodied and substance-based connection between parents and offspring. In the first part of the chapter, I examine several court cases in Europe and North America that have generated intense media interest in recent years, including instances of "mix-ups" at fertility clinics, parent-child gamete donation, and the emerging practice of posthumous reproduction. Through an analysis of these cases, I reveal one important tenet of biological thinking: that the transmission of substance from parents to children connects mothers and fathers in an equal and embodied way to their offspring. I go on to demonstrate that Kamea concepts of sociality offer a compelling foil to the logic of Euro-American representations. Although Kamea draw upon a notion of "one-bloodedness" as a means of describing culturally salient kinds of social relationships, this idiom is not used to designate intergenerational connections; instead, it refers to being the product of the same woman's womb. (Thus, for example, one's mother would be "one-blood" with her own "true" siblings, but not with any person in the ascending or descending generations.) Substance for Kamea lacks the temporal dimension that for Westerners makes reproduction coterminous with genealogical connections. This chapter explores the significance of indigenous ideas, including the extent to which they engender a radically different system of social relationships than that which informs Euro-American kinship configurations.

The central theme of chapter 3 is the human body. I examine how a biological model both draws upon and reproduces an image of the human body as an autonomous and bounded entity that is "complete" in a developmental sense from the moment of conception onward. For Euro-Americans, growth entails the unfolding of a developmental plan that was set in motion at the moment of fertilization.[8] In the first section of this chapter, I illustrate this point by examining the emergence of fetal rights discourse in the West. Drawing upon several cases in which women have been prosecuted for "fetal endangerment," I show that one important concomitant of a biological paradigm is that it constructs an image of mother and child not only as separate individuals, but also as beings with competing rights. In the second half of this chapter, I compare this system of ideas with Kamea conceptions of bodily integrity, wherein bodies are not seen to exist in exclusively bounded terms, but pass back and forth between singular and composite states. I illustrate this point through an examination of the men's cult. More specifically, I show that the aim of male initiation is to separate what is understood to be a conjoined entity—the shared corporeal existence of a mother and her son. Far from being a natural given, the capacity to act as an autonomous agent is elicited by Kamea through intentional human effort.

In chapter 4, I argue that a biological model not only influences how Euro-Americans view the beginning of life, it also informs how Westerners perceive death. I begin with a discussion of recent controversies concerning cloning and use this debate to highlight two interrelated ideas that stem from biological thinking. First is the notion that personhood is inextricably linked with a sense of genetic uniqueness; second is the perception that reproduction is characterized by randomness and unpredictability. Replication is both undesirable and unnatural; rather, it is the constant production of "newness" that is valorized in a biological model. These ideas are then contrasted with Kamea understandings of death, which challenge the nonrecursive approach to life that follows from a biological paradigm. Unlike Europe and North America, where the production of unending novelty and diversity is something to be celebrated, for Kamea it is the ongoing replication of certain identities and relationships that carries sociality forward. By examining Kamea mortuary practices I reveal the existence of an ongoing dialectic whereby relationships based on siblingship and affinity, and the living and the dead, eventuate from one another. This chapter thus establishes that for Kamea, the production of social life rests not on the production of unique individuals, but rather, on the ongoing elicitation of particular sets of relationships.

Chapter 5 examines the social and political implications of extending biology as a worldview to other parts of the globe. I consider this theme by examining two interrelated movements that took place during the final phase of my research: the patenting of New Guinea blood types and the spread of global conservation measures into the highlands of Papua New Guinea. I argue that an important shift is taking place in terms of how Europeans and North Americans conceptualize Third and Fourth World peoples. Increasingly, non-Western peoples are (re-)presented to Western audiences as being "just like us" (i.e., Euro-Americans), in terms of their unwavering commitment to a capitalist economy; they are said to differ primarily in their underlying genetic makeup. This chapter examines the social and ethical implications of these developments, including what they suggest about shifting Euro-American conceptualizations of the human condition.

The final chapter offers summaries and conclusions. I discuss the significance of the material presented both in terms of its contributions to anthropological theory and with respect to the ethnography of Pacific Island societies. I demonstrate that an examination of Kamea (and potentially other indigenous peoples) broadens current debates in kinship studies and our understanding of the impact of biotechnology on European and North American societies. I highlight the extent to which anthropological theorizing on social life must change if it is seriously to confront the implications of imagining kinship in nonbiological terms.

A NOTE ON FIELDWORK AND METHODOLOGY

This study is inherently comparative in nature. My intent is to use Kamea viewpoints as a lens through which to analyze certain assumptions that flow from and inform a biological paradigm. An analytic strategy that relies so heavily on juxtapositional exercises is not without problems of method and interpretation. One of the most obvious shortcomings of this study is its reliance on reified dualisms: the opposition between "Western" versus "Kamea" is the most blatant example, followed closely by "biological" versus "nonbiological." There are many different kinds of people who live in the West. There are also enormous differences among the countries themselves (Bouquet 1993), not to mention salient differences based on class, gender, sexual orientation, and the like. A similar set of caveats could be applied to those persons I gloss as "Kamea." Here, discrepancies based on age, gender, and one's familiarity with the expanding cash economy of Papua New Guinea engender important differences in worldview, some of which will be examined in the analysis that follows. Dualisms have been heavily critiqued in recent anthropological writings. Edward Said, in his highly influential work *Orientalism* (1978), argues that the social sciences have magnified difference in their portrayal of non-Western social realities, and have exaggerated coherence in what are otherwise flexible categories (Bledsoe 2002: 30). However well placed this critique may be, as a heuristic device objectifications are sometimes necessary. Throughout this text, the terms *Euro-American* and *Western* are used interchangeably to refer to predominantly Anglo understandings of the world, particularly those that came into being during the mid-1800s when the discipline of evolutionary biology began to coalesce. The term *Kamea* is used to refer to those fourteen thousand speakers of the Kapau language who reside in Gulf Province, Papua New Guinea, and who are governed from the district office at Kaintiba.[9]

This book relies on a combination of research methods—ethnographic fieldwork and text-based analysis. The sources I draw upon in my portrayal of Euro-American cultural logic are varied, ranging from academic and scholarly accounts to popular print media. Innovations in biotechnology make headline news on a daily basis. Academic accounts that take up the societal implications of these developments have become one of the most vigorous and productive areas of contemporary scholarly research. This has been both a blessing and a curse in writing this book. On the one hand, there has been no dearth of materials to consult. On the other hand, because biotechnology has garnered such intense interest, the debates surrounding its use have become incredibly nuanced in an incredibly short period of time. Because there are many positions taken on any one issue, my summary of these debates is inevitably partial, and at times will undoubtedly strike the reader as being trite. What I have tried to do in this work is to highlight some of the main points of contention, rather than exhaustively inventory the various stances that surround any one controversy.

Fieldwork in Papua New Guinea was carried out over a thirty-one-month period from August 1989 to February 1992. The bulk of my research was undertaken in the central Kamea region at a village named Titamnga, which is situated in Gulf Province, approximately one day's walk from the Gulf Province–Morobe Province border. Culturally and linguistically, Kamea belong to the Angan group, who are perhaps best known to anthropological audiences through the ethnographic writings of Maurice Godelier (1982, 1986) and Gilbert Herdt (1981, 1987).[10] (See map 1.) At the time of my fieldwork, most of the anthropological work that had been carried out among Angan peoples dealt with those groups who resided in the

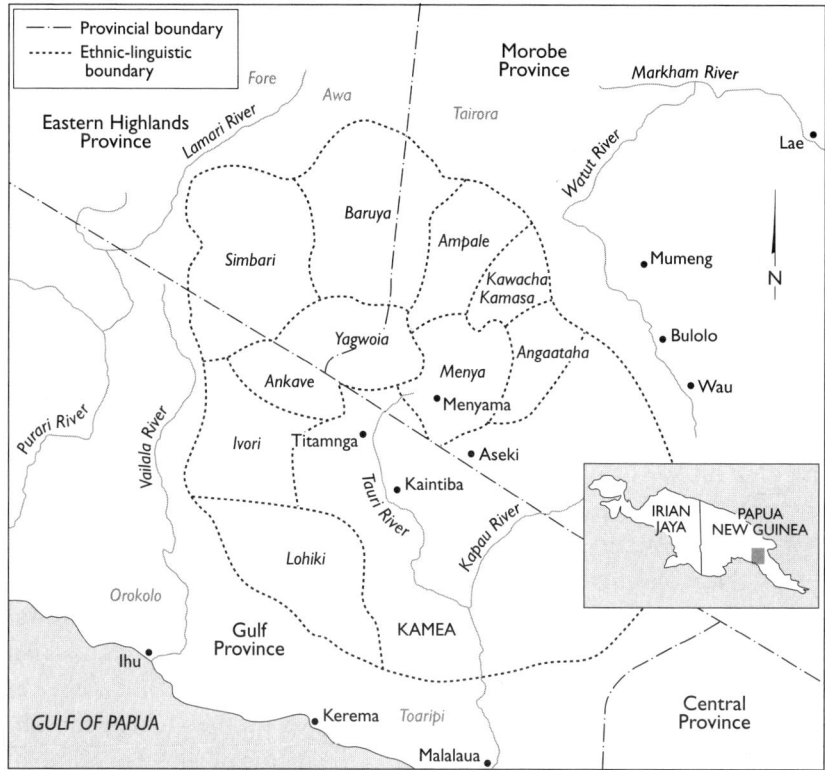

Map 1. Kamea and their Angan neighbors.

far north of their territory (i.e., Baruya, Simbari, and Yagwoia speakers). For this reason, I decided to concentrate my own efforts further south in the hopes that my project would yield some interesting comparative material.

When I first arrived at Titamnga, my command of tok pisin was halting at best.[11] Although I had taken a course in Neo-Melanesian before leaving the United States, everything that I learned seemed to fly from my head when I heard the language spoken at a normal conversational speed. To complicate matters, I soon discovered that a sizable percentage of the local population was as much in the dark as I with respect to the lingua franca of the country. Although most men could speak at least a smattering of Pidgin that they had learned as a consequence of their experience as contract wage laborers on the coast, only one woman—a teenager named Laleto—had any measure of competence in the language. Over my first few weeks, largely through the tutelage of Laleto and several young men, my faltering command of Pidgin was replaced by a fluency, such that when I left Tita-mnga for my first field break nearly four months later, I felt almost as comfortable using Pidgin as I did English. As my own command of Pidgin grew, so too did that of the local women who seemed almost as eager to learn about me as I was about them. Their grasp of Pidgin emerged in tandem with my own—a testimony not only to their incredible warmth, but also their unfailing ability to make a stranger feel comfortable in their midst. After a year or so, I could carry out conversations in <u>tok pisin</u> with most of the younger women, and many of the older ones could understand the language, even if they felt uncomfortable speaking it. Having mastered Pidgin, I set about learning Kapau, which proved to be far more challenging. Shortly after arriving at Titamnga, I made arrangements with one of the younger men—Tarmin Miapeo—to work with me on a regular basis as a language tutor. Through his patient guidance (and the informal teachings of nearly everyone else), I gradually began to make inroads into Kapau, although I never attained the same level of fluency in the local vernacular as I did in Neo-Melanesian.

Members of the academy once entertained the idea that female anthropologists were in a particularly good position to work with women in other societies because their "shared experiences" as members of the same sex would allow them to empathize in a special way with their female consultants. I harbor no such illusions regarding the nature of this research. Nor do I understand gender in the essentialist terms that this position presumes. What I do believe, however, is that local understandings of gender influenced the tenor of my research in several important ways. Certain avenues of knowledge were open to me while others were closed. The greatest difficulty I am aware that I faced concerned speaking with people about male initiation (see chapter 3). Maurice Godelier (1986) has argued that among the neighboring Baruya, differential access to knowledge serves as an important vehicle through which Baruya men establish and legitimate control over women. This statement does not describe the situation that I encountered with Kamea. Among the people with whom I worked, knowledge is not a weapon in the "war of the sexes," but rather, a tool that is used to preserve one's health and well-being. Kamea understand knowledge to have corporeal effects. It follows that certain types of knowledge are appropriate for certain categories of people. In the wrong hands, knowledge becomes a poison—a toxin—that can bring sickness, even death, to the unwary bearer. Restricting access is not a political act, but rather, a medicinal one.

Finally, I should emphasize that this study is written from the vantage point of the central Kamea region; where I lived and worked at Titamnga, based on what I learned from those people whose kindness and patience allowed me to catch a glimpse—however partial—of their world. The Kapau-Kamea occupy approximately two thousand square miles, distributed nearly evenly between Gulf and Morobe Provinces. Prior to settling at Titamnga, I traveled extensively throughout the Kapau-Kamea region and became aware of social and geographical differences. For this reason, I cannot say that the material presented in this account would hold true of all Kapau-speaking Angans, or that it is peculiar to them. I can only hope that it reflects something of the interpretative framework that allows them to interact meaningfully with one another and, to a lesser degree, the anthropologist they graciously welcomed into their midst.

ONE

Cultural Landscapes

MONSANTO MEETS MARY DOUGLAS

August 26, 2001, was a sunny and warm day in southern France. Shortly after noon, a group of men and women began to convene on a small plot of land near the rural town of Auch, a region well known for its culinary delights and picturesque beauty. A distinctly festive atmosphere prevailed. Many arrived on the scene with picnic baskets brimming with Roquefort cheese, foie gras, potted duck, and other regional specialties. To look at the smiling faces in the crowd, one would think it was a belated Bastille Day celebration, rather than the inauguration of a clandestine operation that by the end of the day would render all participants guilty of "trespassing, property destruction, and theft" (Ford 2001). When the assembly reached approximately 150 strong, the small contingent got to work. Armed with scythes, machetes, scissors, and pruning hooks, the group set about razing a plot of genetically modified maize. Within five minutes, the eighty-square-yard plot of corn had been felled, leaving behind nothing but rows of stalks. The protestors then piled the offending crop into the trunks of their cars before driving to the nearby site of Cleon d'Andran, where they cut down another field of experimental corn while police officers watched from the sidelines (Mallet 2001).

What took place on that warm summer day was not the first—or the last—attack on French soil directed against the expanding use of genetic engineering in agriculture. In 1997, a group of protestors in Nerac, also in southern France, demolished a stock of genetically modified corn seed belonging to the Swiss multinational Novataris AG (Egan 2003; Sciclino 2002). Two years later, protestors razed genetically altered rice plants and related research facilities at Cirad, an internationally funded agricultural research institute attached to Montpellier University (Graham 2001; Sciclino 2002).

The driving force behind these and similar campaigns has been the Confederation Paysanne (CP), a militant French farmers' organization first formed in 1987. Led by José Bove, a Parisian intellectual turned activist farmer (Klee 1999), the Confederation Paysanne is seeking to obliterate all genetic crop experiments in France and to protest what it sees as the growing imperialism of multinational biotech companies (Godoy 2003).[1] For his efforts in masterminding these and other demonstrations, Bove was sentenced to ten months in jail, but he served only five weeks. He emerged from imprisonment as one of France's most popular heroes and within days appeared onstage in front of two hundred thousand supporters, where he once again spoke out against what he called "the seeds of death" (Henley 2003).[2] "The judge did us a great service by throwing me in jail," Bove said. "We couldn't have asked for better publicity" (quoted in Sancton 1999).

Since genetically modified organisms (GMOs) first appeared on the scene in the early 1990s, their use has created a storm of controversy. Recent developments in science and technology have made it possible for genes to be transferred from one species to another, either by inserting them directly into the cells of a recipient organism, or by infecting cells with them using an altered virus or synthetic vector. The tools needed to achieve these transfers of genes have been available since the advent of recombinant DNA (rDNA) technology in the 1980s. This technology splices a gene from one organism into a piece of DNA from a virus or some other small object, and then uses that object to bring the new genetic material to a desired chromosome, often the chromosome of another species (Carlson 2001). In this manner, genes from bacteria (in particular, *Bacillus thuringiensis*, or Bt) have been transferred to plants, including corn and cotton, to protect them from a variety of pests.[3] Similarly, genes can be moved between human beings and fish or between animals and vegetables, depending on the desired objective of the transfer (Mulugu 1998).

The first plant food derived from a genetically modified crop to be sold in North America was the FlavrSavr tomato. It was introduced in 1994 and carried a gene that reduced the production of the fruit's ripening enzyme, thereby extending its shelf life. Soon to follow were crush-resistant zucchini, virus-resistant watermelons, and insect-resistant corn and potatoes (Rugg 2002). Crops that tolerate herbicides, such as soybeans and canola, appeared in 1996. These crops tolerate chemical herbicides, allowing farmers to kill early-season weeds in their fields, thereby reducing the need for spraying later (Guyan 1999). The introduction of "golden rice" in 2000 was the first example of a genetically modified crop that was intended to benefit not just its farmers, but also the consumers who eat it. In this case, the consumers include at least a million children who die each year, and an additional 350,000 who go blind, because of a vitamin A deficiency. By adding two plant genes and one bacterial gene to the crop, this "miracle rice" allows beta carotene to be synthesized in the edible portion of the plant, rather than in its leaves, thereby substantially improving the nutritional value of the crop (Kohl 2001; Nash 2000; Ruggs 2002).

At the time of this writing, manufacturers had highlighted six purported uses of GMO technology (de Gruchy 2002):

- To increase crop yields
- To produce crops that can withstand environmental pressures such as drought, excessive soil salinity, or frost
- To increase the nutritional value of plants so that a variety of crops would carry certain amino acids that they currently lack
- To enhance resistance to disease, weeds, and pests
- To reduce the need for fertilizers and other chemicals
- To improve the flavor and shelf life of crops

To date, more than forty genetically modified crops, including chicory, papayas, potatoes, squash, sugar, beets, and tomatoes, have been approved for commercial use by various agencies in the United States (Toner 2002). Researchers are hoping to develop crops in the future that will produce inedible commodities. Two such possibilities include plants that can be used as raw material for biodegradable plastics and plants that can replace animal and petroleum products as the basis of cosmetics (Kenward 1994).

Given all of these decidedly beneficent attributes, one may wonder why anyone would object to the use of GMOs in crop development. Yet object they have. In 1998, farmers in India torched test plots of genetically modified cotton in a demonstration they called Operation Cremation Monsanto (Anonymous 1998; Desmarais 2002). Later that same year, protestors in Ireland destroyed several experimental fields of genetically engineered potatoes (Ivins 1999). In 1999, demonstrators dumped four tons of modified soya beans outside British prime minister Tony Blair's residence at 10 Downing Street. At the 2001 World Social Summit in Port Alegro, Brazil, French, Basque, and Indonesian farmers joined forces with their Brazilian counterparts, uprooting three hectares of GM soya and occupying the laboratories and stores where the seeds were held (Guyan 1999). Although North Americans have tended to be somewhat more accepting of GMOs, here too, the opposition has been spreading.[4] In April 2001, the National Farmers Union in Canada and the National Family Farm Coalition in the United States announced they were exploring joint actions to ban the introduction of GM wheat in North America (Desmarais 2002).

Academic discussions concerning the negative response to GMOs have often focused on safety issues and political concerns. Given that genetic engineering alters the chemical structure of plants, some have argued that these organisms could pose a threat to people with sensitive immune systems, for example by causing unanticipated allergic reactions (Kaiser 2000: 1867). Other commentators have expressed concern about a system that in creating patents (that is, patenting modified food), will also create monopolies, not to mention force "developing" countries into greater dependency on multinational corporations (Pottage n.d. a, n.d. b; Shiva 2000). These arguments have a great deal of force behind them. What comes as something of a surprise is that neither of these concerns appear to weigh heavily on the popular imagination. In September 2003, the British government, under the auspices of the Agriculture and Environment Biotechnology Commission (AEBC), carried out a national survey of public attitudes toward GMOs. Respondents were asked to state their views on genetic modification and the growing of genetically modified crops in the United Kingdom. Did they want genetic modification and genetically modified crops to be adopted in the United Kingdom? If they were prepared to consider the widespread use of GMOs in the future, what information or evidence would they like to see presented first? (AEBC 2003).

While the majority of respondents (a resounding 86 percent) were vehemently opposed to the use of GMOs under any circumstances, human health risks and political concerns rated last and second to last on respondents' lists of objections (AEBC 2003: 19–21). Far more pressing in the minds of many was a vague anxiety concerning the "unintended consequences" that might accompany GMOs (Bremmer 1998). "Right now, GMOs are like a child that is not fully developed," says Micos Ruzicka who runs an organic food club from his home near Prague. "We don't yet know the consequences" (quoted in Becker 2003). This notion of "unintended consequence" is rich in implications. It encapsulates a broad range of ideas concerning how "life" is imagined in the West, including the extent to which a biological paradigm grounds not only Euro-American ideas about "kinship," but also how we imagine our relationship to other beings in the organic world.

One of the most interesting findings of the AEBC report is that the notion of crossing "species boundaries" is what Europeans and North Americans find particularly disquieting. As noted by the authors of this survey, "people opposed to GM were generally far more opposed to trans-species applications, than others, especially GM animals" (AEBC 2003: 22). Central to the folk model of Europeans and North Americans is the view that all constituents of the organic world fall into certain naturally occurring types that can be defined on the basis of their unique reproductive histories. This system of ideas dates back to the seventeenth century, when the botanist John Ray suggested that sexual reproduction be used to classify the natural world. Ray argued that however much variation existed among individuals of the same general type, "one species never springs from the seed of another or vice versa" (quoted in Mayr 1982: 257). Since that time, species have been defined by Western audiences in terms of their ability to mate and produce reproductively viable offspring.

For many Europeans and North Americans, the most disturbing thing about genetic engineering is that it involves taking genes from one organism and introducing them into the reproductive cycle of another, thereby confounding what is seen to be the "purity" of a natural type. The concept of "genetic pollution"—an idea that figures prominently in the anti-GM literature—is grounded in such an interpretative framework. A public-interest story published in the *Progressive* by Ben Lilliston outlines what such a notion of "contamination" entails: "Susan Fitzgerald and her husband operate a 1,800 acre farm outside Hancock, Minnesota. Last year, Fitzgerald's 100 acres of organic corn showed evidence of *genetic contamination* as did her neighbor's organic corn crop. The pollen had traveled more than 120 feet from another neighbor's farm. Instead of selling her organic corn crop for approximately $4.00 a bushel, she had to sell her crop on the open market for $1.67" (2001: 26; italics added). Lilliston's article goes on to discuss a common concern about GMOs: that genes from a genetically modified organism can be carried to places where they are not supposed to go (see also Pollan 2001). The concept of genetic pollution implies an essentialist view of the world. It assumes that life forms fall into naturally occurring types and that it is only within these types (and not between them) that genetic material should be exchanged.

In the minds of many Europeans and North Americans, crossing species boundaries is not only intuitively "wrong," but potentially dangerous as well, insomuch as it "interferes with nature" (Lofstedt 2003). In 2002, the United States Food and Drug Administration commissioned a team of interdisciplinary researchers to examine the safety of applying biotechnology to animal products used for food. After considering the issue, the commission reported that although it had some reservations about the safety of food derived from gene-altered animals, its principal worry when it came to GMOs lay elsewhere. As reported in the *New York Times:* "The 12-member committee of scientists, doctors and other experts said its biggest concern about the new technology was the potential of certain genetically engineered organisms to escape and reproduce in the natural environment. Modified insects, fish, shellfish and other animals could easily escape and threaten their natural counterparts. . . . The panel said gene-altered salmon given the ability to grow at an accelerated rate might compete more successfully for food and mates than natural varieties, causing wild salmon to die out" (Leary 2002). GMOs were created in the laboratory and through intentional human effort; this apparently means that they are no longer "natural." These organisms have been transformed, rendered "unnatural," by the conditions under which they were brought into existence. It follows that GMOs pose a threat to their "wild" counterparts. Should these organisms interbreed with ones that have not been so modified, they will taint what is otherwise a "natural" type. Ironically, the end result of this process of interbreeding will be extinction.[5]

Central to the notion of the "unintended consequence" is a perception of evolution run amok. Moving genes between organisms, it is felt, will lead to evolutionary Armageddon. In the words of Cavanah (2002), a process of "(un)natural selection will take over," giving rise to a world populated by nightmarish organisms and open to life-threatening possibilities. Herbicide-resistant crops will lead to the production of "super-weeds" that human beings will be unable to control in the long run (Goodyear-Smith 2001; Kenward 1994). Pest-resistant crops will enter the food chain and wreak havoc on organisms they were not intended to harm (Knestout 2000; Niles 2001).[6] New diseases will be born, and old ones will mutate and develop resistance to antibiotics that were previously effective (Bremmer 1998). Ultimately at risk, we are told, is the biodiversity of the planet, and with it the economies of human beings the world over. In describing the discovery of pest-resistant corn varieties in Mexico, journalist Michael Pollan writes:

> The country where corn was probably first domesticated, Mexico, is today the source of the crop's greatest genetic diversity. Now that diversity could well be threatened. . . . The presence of transgenes in what some experts call "the cradle of corn" represents a threat to the crop's biodiversity. Should the traits introduced into Mexican fields confer an evolutionary biological advantage (for insect resistance, say) on certain plants, their offspring could crowd out older varieties, leading to the extinction of genes we may some day need. For whenever a good crop suffers a catastrophic failure—as when blight destroyed the potato crop in Ireland—breeders return to that crop's center of diversity to find genes for resistance. Next time around, the genes may be nowhere to be found. (Pollan 2001: 74)

The symbolism manifest in current demonstrations against GMOs is illustrative of the aforementioned themes. It has become commonplace for opponents of biotechnology to label GM crops "Frankenfoods," evoking Mary Wollstonecraft Shelley's *Frankenstein,* in which life is created in the laboratory, only to be set loose on society with disastrous consequences. A similar message underlies the gift of a three-sleeved T-shirt presented to Prime Minister Tony Blair upon his return from a state visit to Barbados in August 2003. Manufactured in Wales by the Cardigan Bay company, "the T-shirt was intended as a tongue-in-cheek critique on the spread of GM in [Britain] against the wishes of the people" (Brindley 2003). Ninety additional garments were made and sent to leaders and policy makers the world over. In the words of Davie Hieatt, one of the company's cofounders: "Part of our company's aim is to make people think about the world we live in and we are trying, in our own little way, to get the issues debated. . . . There are always unintended consequences of the things we do and in 10 or 20 years, no doubt we will find out what they are" (quoted in Brindley 2003).

I have argued that however compelling political and safety-related reasons may be, they do not account for the intensity of public reaction against

GMOs. (To reiterate the findings of the AEBC report, these issues rated *last* and *second to last* on the lists of respondents concerns.) Instead, much of the negative response to GMOs stems from the fact that they transgress what Europeans and North Americans see to be innate distinctions in "nature." As Mary Douglas told us forty years ago, confounding the categories upon which a conceptual system rests will be seen by its adherents as being not only polluting but also dangerous (1966). In Euro-American folk wisdom, essential distinctions exist between certain forms of life, and these, in turn, reflect their unique reproductive histories.

It is highly significant that the practice of mutagenesis by plant breeders has not met with the same kind of resistance that has accompanied the manufacture of transspecies GMOs. Mutagenesis involves exposing plants to a variety of mutagens, including chemicals or gamma radiation, in order to induce mutations in the structure of the plant, thereby giving rise to novel characteristics that can be passed on to subsequent generations (Davies 2001: 427; cf. Nuttall 1998). The new traits can then be selected and introduced into new cultigens using traditional plant-breeding techniques. The practice of mutagenesis has not prompted any strong reaction on the part of the public despite the fact that it also uses biotechnology to intentionally introduce a change in the genetic structure of an organism. Unlike those techniques of genetic engineering that rely on transgenic (i.e., "interspecies") applications, mutagenesis does not involve crossing so-called species barriers (Davies 2001: 427). Consequently it is viewed as nonthreatening.

In the remainder of this chapter, I will sketch the contours of a very different world, one in which biological reproduction is not seen to ground a set of fixed and enduring distinctions. If "crossing" species boundaries is threatening to the social order in the West, for Kamea, by contrast, it is constitutive of it.[7] For Kamea, human reproductive potential is *directly tied* to other "species" rather than being defined in contradistinction to them. As we shall see, this carries with it a number of important implications. Life, generation, and property relations all take on a radically different slant when viewed through Kamea eyes. A second aim of this chapter is to highlight the extent to which a particular vision of biology has structured more than Euro-American conceptions of the "natural world": it also has entered social science thought, influencing how anthropologists interpret the social life of others. Before considering Kamea human-environmental relations in detail, I turn briefly to the world of contemporary anthropological theory.

Over the last few years, the discipline of anthropology has witnessed a resurgence of interest in matters pertaining to the "physical" world. Indeed, the related themes of environmentalism (Brosius 1997, 1999a, 1999b; Escobar 1998a, 1999; Milton 1993; Zerner 1994, 2000) and "historical ecology" (Crumley 1994) are fast becoming common concerns to the practitioners of our discipline. Although an interest in "nature" is hardly new to an intellectual enterprise that defines itself as being oriented toward the study of its supposed antithesis—"culture"—this new brand of theorizing is claiming to go where none of its analytic forbearers have gone before. The cerebralism of structural analysis (Douglas 1966; E. Leach 1964; Lévi-Strauss 1966, 1967) and the adaptive functionalism of cultural ecology (Damas 1969; Hardesty 1977; Lee and DeVore 1968; Rappaport 1968, 1979; Vayda 1969) have been replaced with a view of the "natural" world that is eminently social (or better yet, "socialized") in nature.

One of the most cogent statements to emerge from this new theoretical vein is set forth by Philippe Descola in a paper entitled, appropriately enough, "Societies of Nature and the Nature of Society" (1992). Motivated by a desire to "isolate unconscious schemes of praxis through which each society objectifies certain types of relationships with its environment" (Descola and Pálsson 1996: 17), Descola argues that "there is a homology between the way in which we address "nature" and the way in which we address 'others' " (Descola 1992: 111; cf. Bird-David 1993). Drawing upon both published accounts and his own research in Amazonia, Descola suggests that human beings draw upon the categories of social life to structure in conceptual terms the relationships between themselves and "natural" species (Descola 1992: 114).[8] In the South American rain forest, this plays itself out via patterns of marital alliance. Where women are obtained from enemy groups on head-hunting raids, human beings are said to perceive their link with "nature" as being based on predation. But where women circulate between exogamous groups through a principle of generalized exchange, human-environmental relations are characterized by an ethos of balanced reciprocity. "Nature," in this view, is neither purely "mental" nor "material" in its range of ramifications; rather, it takes its cue from the particular nexus of social relationships within which it is embedded (cf. Bird-David 1993; Bloch 1992; Ingold 1994; Rival 1993).

Descola's thesis is particularly compelling to consider from the vantage point of highland New Guinea ethnography. From the late 1950s, when anthropologists first began to carry out research in the newly contacted high-lands of Papua New Guinea, the search has been on for a theoretical paradigm that could adequately account for the forms of sociality they encountered. Unlike their colleagues working in Africa and elsewhere, ethnographers of New Guinea have always been hard pressed to identify anything resembling stable social groups. Described in terms of its "looseness" (Pouwer 1960; Watson 1965), "plasticity" (Kaberry 1967), and interminable fluidity, Melanesian sociality appears to fly in the face of everything that Westerners hold dear about an ordered society. The difficulties that ethnographers have faced were described early on by J. A. Barnes (1962) in a now-classic article in which he documents the inadequacy of descent-based models for this part of the world. New Guineans, in short, appear to be far less interested in building a "society" than they are in creating those distinctions that make particular types of social relationships possible (cf. M. Strathern 1988; Wagner 1975).

Given this, an interesting problem presents itself: if, as Descola argues, human beings model their relationships with "nature" on how they treat the "other," what happens when "otherness" is subject to ongoing renegotiation—when sociality is a *becoming* rather than a *fait accompli* and cultural distinctions must be *elicited* rather than merely acted out?[9]

In the following sections of this chapter, I consider these issues from the perspective of Kamea. A central question I ask is: how do human beings conceptualize their relationship to other life forms when their ties to one another are subject to ongoing creation? Through an examination of how Kamea frame intergenerational continuity, I will show that the resources upon which they depend for a living—the land and the different types of flora and fauna that they utilize—are not simply appropriated via preexisting social ties, but instead furnish an important venue through which salient social distinctions are created in the first place. Gender and different categories of social relationships sediment out of the differential uses to which the non-human world is put. What Europeans and North Americans separate as the discrete domains of "nature" and "culture" are for them part of an encompassing social whole. The pages that follow thus establish ethnographically the mutual embeddedness of human social life and the organic world.

In pursuing these themes, I turn first to an examination of how Kamea occupy space and how they live upon the serrated landscape they call home. Here, we shall see how indigenous conceptions of male (*oka*) and female (*apaka*) go together to form a productive whole at the level of the household. I then turn to a more detailed examination of how the landscape is perceived, including its role in defining human social relations. I will demonstrate that lineal relations (what an earlier generation of anthropologists glossed as relationships based on "descent") are created among Kamea as part of an ongoing process whereby men come to inscribe their social identities on the land. Through the planting of select trees (including several varieties of pandanus, ficus, and palm lilly), intergenerational links are formed between men who work the same land, not on the basis of shared biogenetic substance, but rather through their joint ability to share in the creation of a personified landscape. This chapter then turns to a detailed consideration of the connection between maleness, intergenerational time, and the organic world. In the next chapter, I pick up on the other half of this "productive whole" by considering how relationships defined through women intersect with male sociality and the ties that men establish with a variety of environmental forms.

CULTIVATING CONNECTIONS

An outsider approaching the Kamea habitat for the first time is likely to be struck by an overwhelming sense of isolation. Miles of unbroken rain forest dominate a landscape that appears to lie cloaked beneath a perpetual blanket of low-lying fog. Far removed from vehicular roads and the quiet bustle of the larger coastal centers, the Kamea territory is a land of visual and topographical extremes. Containing nearly two thousand square miles of bush, it extends from the softly rolling hills of southern Gulf Province through the razor-backed ridges that make up New Guinea's central cordillera. Population densities throughout the region are among the lowest in the country, averaging in some places fewer than four persons per square mile (Provincial Data System Community Register 1978).

Across this vast expanse of territory, the household emerges as the operational social unit. Although mission and government agents have been strenuous in their attempts to promote a collective village-based lifestyle, Kamea have always exhibited a marked preference for living in the bush. It is common to find that even when a couple maintains a house in one of the newly created villages, they are likely to spend a considerable portion of their time in the bush, sleeping in garden houses for days, sometimes even weeks. For some, this is a matter of practicality. Gardens are often situated several hours' walk from settlements, making it difficult to return to the vil-

Figure 1. A mountain airstrip. In the absence of vehicular roads, one travels through the Kamea region either by foot or by small aircraft. Photo by Sandra Bamford.

lage at the end of each day's work. But more than simple pragmatism is involved. Bush living is a way of life. Indeed, prior to the imposition of colonial rule during the 1960s, it was the only way of life; people lived in solitary homesteads scattered throughout the bush, much to the chagrin of patrol officers, whose task it became to see to their administration. For many people, this pattern continues to be the preferred mode of residence. After spending a day or two at Titamnga, most people are ready to seek out the quiet solitude of their gardens, to escape the watchful eyes of their friends and neighbors and the ongoing demands of social reciprocity that living in close quarters frequently entails.

A man, his wife, and their children, then, form a residential unit. The permanent men's house—a feature commonly found elsewhere in the highlands (Godelier 1986; Herdt 1981; Jorgensen 1981; Poole 1981; Read 1952, 1954)—is notably absent throughout the Kamea area. Occasionally, two brothers will build their houses side by side, or a son will settle near his father, thereby forming a tiny hamlet. In the past, these settlements were staked off by rows of cordyline (palm lily) and wild sugarcane for defensive purposes.

Like most of their neighbors, Kamea derive a living from a combination of shifting horticulture and the raising of pigs. Gardens are generally cut from secondary forest and are productive for a period of two to three years. The main crop is sweet potato, which is cultivated in family plots at elevations ranging from two thousand to six thousand feet. Sugarcane, banana, taro, and pitpit are planted intermittently throughout the garden, wherever space permits. Because seasonal changes are slight in terms of rainfall and temperature, crops are planted and harvested on an ongoing basis. As the yields of the garden currently under production begin to diminish, a new patch of ground will be cleared and planted in anticipation of future use. Old gardens are allowed to return to secondary growth before being used again.

While men and women cooperate in subsistence activities, there is little collaboration beyond the household level. The pattern frequently encountered in other highland societies, whereby several related kinspersons till gardens together as a team, does not take place in the Kamea region.[10] This is related, in part, to the system of land tenure. Unlike those areas of Papua New Guinea where corporate groups are vested with the joint ownership of a territory, plots at Titamnga are held and farmed by individual families.[11] When a man dies, his holdings are generally distributed among his sons, each of whom manages his share of the estate as he sees fit.

Although most people garden on their own land, some flexibility exists in land-tenure arrangements. Affines, classificatory siblings, and matrilateral kin are often granted temporary rights to use a specific plot of ground. Should a woman find, for example, that she and her husband are always "hungry," she might approach her father or brother with the request that she be allocated space to work a garden on their land. Such appeals are almost never denied. Rights to the borrowed land end with cultivation; once the crops have been harvested, the ground automatically returns to the control of the original owner. At the time of my fieldwork, probably 20 percent of the gardens under production were being worked by persons who lacked permanent title to the land. These nontitled individuals come and go, gardening for a time on land belonging to them and then on land belonging to others. Occasionally, more enduring rights to land are negotiated, a point that I will discuss in greater detail below.

Hunting and gathering follow gardening in terms of economic importance. A wide variety of food is collected from the forest, including several types of edible fruits, greens, and mushrooms. Women and children collect these items, along with several kinds of small game (i.e., insects, birds, frogs, lizards, and rats) as they go about their daily rounds. Men hunt larger game animals, such as wild pigs and marsupials. While these contribute minimally to the diet, they figure centrally within the context of social and ritual presentations. Finally, Kamea also raise three types of animals: pigs, chickens, and dogs, the latter for hunting purposes.

Subsistence activities are structured in such a way that cross-sex relationships are a necessary component of production. Men fell trees and clear the underbrush, build fences and construct the support poles for sugarcane and banana. They are also responsible for the initial firing of plots and for planting the cordylines and stakes that act as boundary markers. Yet, if men are involved in the initial stages of garden work, it is their wives who assume the responsibility of day-to-day production. Kamea women do most of the planting, weeding, and harvesting, and take care of those minor garden repairs that are necessary on a regular basis. The moral character of a woman is frequently judged in terms of her capacity to make a garden flourish. The "good woman" (*awati apaka*) is one who not only maintains an amicable relationship with her husband, but who possesses the knowledge and commitment to make the cultigens in her garden thrive. A man who is angry with his wife often voices his complaints through euphemisms that critique her gardening capabilities.

In its composition, the garden exists as an externalized image of those relationships that went into its production. Just as the household combines male and female in the form of husband and wife, the garden exists as an embodiment of their joint labors. The men and women of Titamnga maintain that a well-balanced garden should contain a healthy mix of both "male" and "female" crops (Herdt 1981; LiPuma 1988; Tuzin 1972). Taro, banana, and sugarcane are all male cultigens par excellence in that they are all said to do particularly well under the ministrations of men. Should a man find himself alone with nightfall approaching, he would not hesitate to harvest any of these crops from his garden, something that he would rarely do with any of the other, "female" cultigens (sweet potatoes, cabbage, pumpkin, onions, tomatoes, corn, and a variety of leafy greens).[12]

It is significant that many of those crops planted by a husband and wife team are spoken of as being the "children" of the married couple. Created through the union of male and female efforts, the garden exists as an objectification of the marital relationship. If the relationship thrives, the garden thrives; if the relationship falters, the garden similarly falters. In terms of indigenous categories, cultivated pandanus nut trees (*anaka*) occupy a particularly compelling place in Kamea thought. Possessing a life span of some sixty or seventy years, these trees reflect the care and attention of several generations of human beings. Pandanus nut trees are "personalized" in the sense that each is assigned its own name, like a child, and the memory of all those who planted and tended to each tree over time is carefully preserved. Land disputes are frequently discussed entirely in terms of the history of these trees, which serve as points of attachment to specific tracts of land.

By engaging in repeated acts of cultivation, traces of human identity come to be left on the landscape. Men and women move through a "mosaic of special places, each stamped by human intention, value, and memory" (Buttimer 1976: 283; cf. M. Kahn 1990; Maschio 1994; Rodman 1987; Schieffelin 1976). Patches of secondary regrowth, former house sites, and tracks through the forest all bear the imprint of human activity. These sites are named, and their names provide a record of human movement across the landscape over time (see Maschio 1994). As Ingold (1993) has noted within a similar context, "the environment tells, or rather *is* a story. It enfolds the lives and times of predecessors who, over the generations have moved around in it and played their part in its formation" (152). Even a young Kamea child possesses detailed knowledge of the multiplicity of names that render the landscape meaningful in human terms. As my consultants taught me some of these names, they recounted stories about the people who had occupied the land before them and of those events that had given rise to a particular place and its name.

Children ease into this world of adult responsibility where the history of human beings and other life forms coalesce. Infants and toddlers of either sex stick close to their mother as she goes about her daily rounds. While women garden, their sons or daughters sleep nearby in the shade of a tree, frequently cradled in a string bag. Those who are capable of doing so frolic about on the sides of slopes, unaware that a world of any consequence exists beyond the narrow confines of the sweet potato patch. Youngsters rarely venture far from the watchful eyes of their mothers. A child left unattended, even for a short period of time, is an easy mark for one of the many categories of bush spirits (*hiey'ya*) who live nearby and are known to steal human children and bring them to their rocky lair. By the age of seven or eight, girls are expected to leave behind the carefree exuberance of youth, and to help their mothers with domestic chores, including the care of younger siblings. Daughters will be allocated a small corner of the family garden where they are expected to plant and tend their own cultigens. A girl who fails to pull her weight in terms of subsistence activities will feel the consequences of her actions at mealtime. Gradually, almost imperceptibly at first, a mother will begin to allocate smaller and smaller portions of food to her daughter in the hopes of encouraging her to tend to her own needs. By the age of ten or so, girls are expected to be largely self-sufficient, with the exception of owning land and meeting those subsistence needs that fall outside the female half of the division of labor.

While girls are taking up the digging stick (*uka*), their brothers are entering into a more mobile way of life. From the age of nine or ten, through the teenage years, young boys take to exploring the high mountain forest. Male youths range over a vast expanse of territory, in the process becoming deeply familiar with the contours of their habitat. They can often be seen heading out in groups of two or three to catch tadpoles in a stream or to sharpen their skills with a bow and arrow in the bush. They may spend a few days visiting kinsmen in salt-making camps, or perusing pandanus groves in search of an unexpected windfall of nuts.

As he gets older, a boy will be shown the land of his father. Gaining access to land is not an automatic concomitant of patrifiliation. Rather, it is contingent upon knowing the history of the land in question and those features of the landscape that mark it as belonging to a particular line of men. It is the responsibility of a boy's father to walk him about the land he may one day inherit, pointing out the boundaries of the allotment and the exact location of different resources (cf. Rodman 1987; Rohatynskyj 1990). A boy will also be taught who worked the land before him, and where his predecessors gardened, built their houses, and hunted for game. Fathers also teach their sons about the mythic events that led to the creation of those topographical features that characterize the present-day terrain.

Sometime between the ages of nine and thirteen, boys are initiated into the men's cult. The building of a temporary men's clubhouse in the bush inaugurates the ritual cycle, and it is here that boys will have their noses pierced by older initiated men during the first phase of the ritual sequence (see chapter 3 for further details). Having been initiated, a young boy is ready to marry and to form a household of his own. Most Kamea marry for the first time in their mid to late teens, following the onset of menarche in girls and the completion of initiation by boys. Second-degree cross-cousins (on either the mother's or father's side) are the preferred category of spouse, and in most cases constitute the actual choice of a marriage partner. Although marriages are arranged by the parents of the bride and groom when their sons and daughters are still quite young (often in their infancy), the actual timing of the event is subject to considerable speculation. The most definitive sign that the nuptials are imminent is when a young man begins to clear a garden of his own for the first time. Once the land has been cleared and is ready for planting, the girl's parents will bring their daughter to her husband-to-be. By this act, the couple is wed and a new garden will be brought into production.

The contours of the lived world are beginning to emerge. An understanding of how men and women move through space and relate to nonhuman forms begins with the household and those relationships of which it is composed. However, men and women at Titamnga do not simply act upon the nonhuman world vis-à-vis an assemblage of ready-made identities and relationships. Salient social distinctions emanate out of the organic world, rather than imposing themselves upon it. To grasp this point, we need to delve deeper into Kamea understandings of the organic world, which is the subject of the next section of this chapter.

NARRATIVES OF CONTINUITY

The relationships that men form with specific tracts of land are remembered through time and form the basis upon which generational continuity is conceived. Drawing upon these links, Kamea construct an account of their own history that extends back in time to the origins of humanity. The following myth establishes the baseline upon which Kamea understandings of the past are built.

Akeanga was the maker of all <u>tambuna</u> (ancestors). He was married to two women who were related as sisters. One day Akeanga went out to look for game in the bush and he returned a short time later with a <u>kuskus</u> (a type of tree kangaroo) in hand. He butchered the animal and carefully wrapped its meat in the bark of a tree, and then handed the bundle to his wives so that they might cook it. Without the women knowing it, however, Akeanga had slipped human feces into the bark container along with the meat.

The two women made a fire and set about cooking the game while their husband covered up with a *malo* (bark cloth) and went to sleep. When the meat was cooked, the younger sister removed the container from the fire and split it open. Upon viewing the contents she immediately became suspicious and said to her elder sister, "Our husband has tried to trick us, this is shit." The two women became very angry. As Akeanga slept, they quietly crept outside and set fire to the house with their husband still inside. The women watched the fire progress until they were certain that it blocked the door, and then headed off toward the bush.

The women walked steadily for about five minutes and then turned around to examine their handiwork. From an adjacent ridge, they could see smoke rising from the burning house as it filled the horizon with a dense black cloud. The women traveled further and then turned to look again. As they watched the fire progress, they could hear a popping noise as their husband's stomach exploded in the fire.

The sisters hid in the bush for one day. They slept alongside a pool of water and in the morning they returned to the scene of the house fire. Using a stick, they carefully broomed up the debris from the fire, separating out the bones of their husband from the rest of the ashes. The women then bundled up the remains of their husband in a bark cloak and deposited his bones inside a pool of water.

With Akeanga safely housed in his watery grave, the sisters continued on to a house in the bush where they slept for three consecutive nights. On the fourth morning, they returned to the pool of water and saw that the bones of their husband had turned into tadpoles. The women went back to the house in the bush, and after some time had elapsed, they returned to the pool of water once again. This time, they saw that the tadpoles had turned into tiny little men who were splashing about in their aqueous habitat. The women went back to the house in the bush where they began to make arm bands (*ituka*), net bags (*ka*), bark cloth (*malo*), grass skirts (*aka*), and the "back apron" (*simga*) traditionally worn by men. When they visited the pool of water several days later, they saw that the men had performed initiation on themselves, and that each now sported a hole in his nasal septum.

The sisters returned to their bush house where they remained a long time. The elder sister set about making grass skirts for men (*apisa*) while the younger sister made several grass skirts for women (*wiwonga*). When they checked the pool of water several days later, the "tadpole-men" were nowhere to be found. Leading away from the pond, stamped in the soft black earth, were a series of footprints which the sisters followed to the base of a tree. As they approached the tree, the women could hear the men talking inside, and they took a stone ax and made an incision in the trunk. Several full-grown men popped out. The sisters made a second slit and the same thing happened. The women then proceeded to make other cuts in the tree and each time they did so another coterie of men emerged.

The women gave names to all the men. Those who emerged from the trunk were called Apea, while the Amdea issued forth from the smaller branches higher up. White men (Europeans) came from holes in the leaves, while men named Ooyena were released from an opening in the roots.

With each new cut, the women called out a different name which they assigned to those men who had issued forth from that particular incision: some men were called Nowtia, some E'oota, some Yakusana, some Tomasa, and so on. The women then presented the men with all of the things they had made—the bark cloth, arm bands, and string bags—and allocated to each named segment a specific tract of ground. Some men were sent to live in the mountains near Menyamya while others went to Pikiango, a stone near Kaintiba. A bridge was built over the Tauri River and some men were sent to live at Kamina and Kanabea. The ancestors of all white men headed toward the coast, taking all of the cargo with them. This is how human beings came to be in the world.

I heard this myth recounted, with minor variations, several times during the course of my research. All accounts described how Akeanga's charred remains came to furnish the basis from which humanity was created. The myth establishes the generative potential of the landscape. Not only was humanity once contained in the trunk of a tree, but through this gestational process, unity (i.e., Akeanga himself) is replicated and differentiated (i.e., he becomes many people with heterogeneous names).

When I first heard this myth, I took it as a charter of rights that could be used to explain how people were attached to specific parcels of land. The myth appeared (at least in my mind) to suggest a segmentary model of society wherein one's position in space could be read as a map of genealogical reality. After hearing this myth, I became confused when people would recite to me complex stories of migration that outlined how they came to reside in the Titamnga area. If in mythic times Amdea men were sent to occupy land to the west of the Tauri River, this meant little in terms of where their descendants held ground today. Autochthonous rights in land were not entertained by the people with whom I worked. What these stories had in common with one another was the idea that long ago, in the quasi-mythic past, human beings were more mobile than they are today. Men did not simply reside where their fathers lived. Instead, they set out to found their own place; where they settled and worked the land is where their "descendants" live today. By clearing new land, building houses, and battling mythic foes rights to use a given tract of ground were established. The myth of origins does not describe the basis of a segmentary land tenure system. Having been released from the tree, people moved out in different directions and *through their work* established claims to land. It is through the process of investing one's self in the land (and having the land, in turn, come to be a part of one's person) that rights to use it are negotiated and renewed.[13]

The myth of origins establishes the starting point from which Kamea construct an understanding of their own history. Melanesianists have long noted that the people with whom they work tend to be rather poor genealogists. For Kamea, it would be more accurate to say that they are not genealogists at all, in that sexual reproduction is not used to frame human social relationships through time (see chapter 2). Although most of the people with whom I spoke could recall the names of their maternal and paternal grandparents when asked, beyond this point ascending links were only hazily remembered. There were, however, some notable exceptions. A few men, generally older and renowned for their wisdom, could recite complex histories (defined along nongenealogical lines) that extended back in time some ten or fifteen generations, at which point they began to blend with the mythical past. These more detailed social histories, known as tambuna storis, commemorate the ties that men form with specific tracts of land.

Tambuna storis outline the comings and goings of particular men; where they traveled, how they lived, who they married, and how the land was transformed through their work. Portrayed as a recollection of actual events, these stories have a perceived historical presence. They generally open with a named ancestor heading out to explore a previously unoccupied tract of land. As the tale unfolds, the narrative assumes the structure of a journey through space where the hero travels from place to place, planting trees, clearing gardens, working houses, and engaging in other types of productive endeavors (Maschio 1994; Munn 1973; Myers 1986). He may stop to make salt at one location, hunt for game at another, and collect stone from a rock quarry at a third. Wherever he goes, he generally manages to leave something of himself behind in the form of making the land more habitable for future generations.

Tambuna storis are cited in legal disputes and more specifically, in litigation over titles to land. Kamea say that men are expected to "follow in the footsteps of their father." Where their tambuna walked about, cleared a garden, and made a house is where they and their sons should do the same. Such a pattern of land use establishes a continuous line of cultivation extending from the mythic past to the present. Knowledge of this history, of one's ties to place, is of critical importance in establishing claims to land (cf. Rodman 1987). Some men become particularly conversant with respect to those ties that bind them to place. However many others lack the knowledge necessary to defend their claims in the face of competition from rivals who either know more, or who can speak more persuasively on their own behalf. Within the context of the contemporary setting, those men who have spent time working on the coast or in the goldfields at Wau are at a particular disadvantage. They return to Titamnga with cash in hand, but have little knowledge of those links that bind them to place. These ex–wage laborers are constantly in danger of losing access to their father's land. The final result is the figure of the "rubbish man" (rabis), the distant matrilateral relative who goes from place to place borrowing garden land and living off of the charity of others (Rohatynskyj 1990).

Yet, knowledge alone is not enough to establish rights to land. An individual who fails to *activate* these claims through the investment of his own labor in the land is apt to break the continuous male line. Not only will his own rights be called into question, so too will the claims of his sons.

What one sees, here, in terms of land use, is the ongoing elicitation of social relations. Using land and moving through space is not only the performance of sociality but also the means whereby ties are created between men through time. While rights to use land typically devolve from a father to his sons, this pattern of transmission is neither necessary nor automatic. A man who lacks male offspring can allocate his holdings to any other male. So long as the recipient uses the land in question on a continuing basis, his rights to it will not be called into question. Similarly, existing rights can be rendered null and void if an individual fails to use the land in an appropriate manner. The situation I will now describe is a case in point.

Not long after I arrived at Titamnga, I became acquainted with Netsap, a man in his mid to late twenties who was married and lived with his wife and infant daughter. Netsap occupied a house that was situated close to my own, and I would often see him heading off to the bush early in the morning with a slingshot or bow and arrow tucked loosely under his arm. From the very beginning, Netsap struck me as an unusual figure. In the first place, I could find no one (with the exception of his wife) who would admit to having a close personal relationship with him. He was consistently left out of the social histories that I collected and was almost never addressed by anyone via the categories of Kamea kinship terminology. Netsap was also different in terms of his mannerisms. Most Kamea men that I knew exuded an unmistakable aura of self-confidence; they were self-possessed, quick to anger, and aggressive in their personal style. Netsap, by contrast, possessed none of these characteristics. He was shy, self-effacing, and almost never offered up an opinion of his own during the frequent and often animated village courts. Even young children spoke about him in a markedly disparaging manner and were able to assert with a surprising measure of confidence that he would never amount to anything. Netsap, it seemed, had fallen through the cracks of conventional sociality.

Roughly six months into my fieldwork, I discovered that unlike all other married men in Titamnga, Netsap had no land to call his own. Despite the fact that he was an as ples man, that is, he had been born at Titamnga and lived there his entire life, he had been effectively disenfranchised from the system of land ownership. Possessing no land meant more than losing the "means of production"; it also contributed to the inability of other men and women in the community to define Netsap in culturally meaningful ways. On the basis of numerous discussions with other men and women, the following story of Netsap's circumstances gradually emerged.

When Netsap was only a few years old his father, Gonapawi, passed away. His mother subsequently married another man from Titamnga and Netsap grew up in the house of this mother's second husband. Taken by itself, this is hardly an unusual happening; many Kamea boys are raised by men other than their fathers. What makes Netsap's situation somewhat unusual by Kamea standards is the impact that Gonapawi's death had on the life of his son.

As previously noted, gaining access to land is contingent on knowing the history of the land in question and the relevant features of the landscape that mark it as belonging to a particular male line. A boy's father will typically take him around and show him the land in question and teach him the history of its cultivation. Netsap's father died before he could transmit these experiences to his son, and consequently, Netsap was in a tenuous position when it came to asserting his own rights to land.

Several people from Titamnga told me that under normal circumstances, a boy's "second father," that is, his father's brother, would come to the aid of his sibling's son. In the case of Netsap this never took place, and the question of why it never took place is crucial to the discussion at hand. When Gonapawi died, one of his brothers attempted to claim his widow as a second wife, but Inema (Netsap's mother) chose to marry someone else. An argument ensued that led to bitter feelings all around, and consequently none of Netsap's paternal kinsmen took it upon themselves to protect the interests of the child. Netsap survives today by being granted temporary rights to use other people's land (generally land belonging to his wife's kinsmen), although the land must be returned to the original user each time the crop is harvested. He is then forced to find another relative who might be willing to lend him land. Should Netsap have a son of his own one day, the boy will similarly be forced to make his own arrangements with respect to negotiating rights to land.

As can be seen from the foregoing, gaining access to land is not a simple by-product of patrifiliation. That Netsap was Gonapawi's son was not subject to debate, but this did not guarantee that he had any rights to land.

Most accounts of Netsap's circumstances differed only in minor detail. However, one of my consultants did add an illuminating twist to the tale. Gonapawi, I was told, led something of a transient existence. Instead of investing his own labor in the land by working a garden and planting trees, he spent his days roving through the bush in search of game and uncultivated plant food. As one man explained: "Gonapawi was 'public.' He had no ground. He simply wandered from place to place. To have ground, you have to stay in one place and plant things. That is how we got our land. Netsap's father simply wandered so he had no land. Now that his child has grown up, he is experiencing difficulties." By not impressing something of himself onto the land, Netsap's father failed to become part of the humanized landscape. Several years later, his son lacked any connection to the land and perhaps even more significantly, he had only tenuous attachments to other people in the community. Over time, neither Netsap nor Gonapawi will figure in <u>tambuna storis</u> insomuch as these tales are a history of human connections to the environment. Both men will, in effect, cease to exist in intergenerational time.

In contrast to the genealogical model of Europe and North America, the men and women of Titamnga construct lineality retroactively, by telling <u>tambuna storis</u> that commemorate men's investment of themselves in the landscape. Corporate descent groups and abstract rules governing the inheritance of property mean little in terms of who gets remembered as <u>tambuna</u>. Lineal continuity is not coterminous with Western ideas of descent. Indeed, in many ways it makes more sense to speak of relationships of *ascent*, in that intentional human effort is required to attach oneself to the male line. Male relationships have little permanence in the history or memory of the people without having been objectified in the land.

EMBODIED CONNECTIONS

The ability to "read" the landscape is central to Kamea ways of being in the world. From the perspective of any particular person, the environment is imbued with social significance. Yet, more than a process of objectification is involved. To view the landscape as a template upon which human social life is inscribed is to address only half the picture. For Kamea, moving through space is also an act of incorporation whereby the landscape comes to be embodied by human beings in meaningful ways. By engaging (or alternatively, not engaging) in specific acts of consumption, components of the nonhuman world enter into the constitution of persons. It is, in part, through such acts of incorporation that Kamea come to define themselves as gendered beings with the capacity to form specific types of social relations.

One of the most important secrets revealed to boys at the time of their initiation (see chapter 3) concerns the use of the *yangwa* tree, a type of ficus. Scattered at regular intervals throughout the Kamea habitat, this tree is a pervasive feature of the landscape. In addition to its publicly acknowledged use in the production of bark cloth (*malo*), the *yangwa* tree has a more compelling use that is known only to initiated men. The inner core of this tree contains a milky white sap that, if drunk, is believed to replace semen which has been lost through sexual intercourse. As one man explained: "If you want to come up 'new' (nupela) again, you will go to the forest without women. You will get the juice of this tree and you will drink it. If a man wants to 'make work' (i.e., engage in sexual intercourse), this juice will change him. The tree has an unlimited supply of milk. If a man drinks it, he will come up 'new' again."

A married man will begin to plant *yangwa* trees shortly after the birth of a son, using cuttings that have been taken from trees that he acquired from his own father. These trees mature slowly; ten to fifteen years is required before their bark or their sap is ready to be used. As the tree grows to maturity, its development is seen to parallel the growth of the child himself. By the time that the tree is ready to be used by the son, the latter will have taken a wife himself, and he will begin to plant *yangwa* trees in anticipation of his own son's future use.

The cultivation of *yangwa* trees establishes a link between men of different generations who are not otherwise united by bonds of shared bodily substance. Although physiological reproduction is not emphasized as a point of connection between parents and children, men nonetheless provide their sons with an externalized and partible equivalent of seminal fluid that gets handed down through time and across generations. Men plant trees, and trees, in turn, elicit the growth of future generations of men. Thus, the landscape comes to be embodied by men in such a way that it provides them with a continuing aspect of male identity that persists through time.

I began this chapter by considering recent controversies surrounding the use of genetically modified organisms. I argued that Westerners' widespread refusal to accept these forms stems, at least in part, from the fact that they challenge many cherished assumptions about the world, particularly the idea that sexual reproduction grounds a series of fixed and primordial distinctions in the "natural" world.

There is a profound irony associated with this position, in that it relies on an essentialist view of the world that is very much at odds with contemporary evolutionary theory. One of the most significant contributions that Darwin (1979) made in the *Origin of Species* was to challenge the preexisting Platonic vision of the organic world in which species were fixed, eternal, and universal—a supposed reflection of God's divine handiwork in creation. *The Origin of Species* was concerned less with explaining species' origins than it was with destabilizing the category of species itself. In Darwin's hands, one could talk about species only as a statistical abstraction (Mayr 1976: 27). The organic world was not characterized by sharp discontinuities, but instead by fluidity, flux, and gradation.

A foundational tenet of Euro-American thought is that physiological traits, along with the social relationships that often accompany them, are based on a principle of "descent." Characteristics defined on a biogenetic basis are "passed down" through the generations forming an irrevocable bond between parents and children. It follows that contained within the bodies of living human beings is a protracted history of procreative events extending back in time from the present to the remote past.

This premise carries with it several important implications. Perhaps most significantly, it precipitates a cleavage between human beings and other life forms. While human beings are linked to other species on the basis of common descent, they are separated on the basis of their current reproductive history. It follows that other life forms play no role in defining human social relations; nor do they enter into the constitution of persons. Land can be lived upon. It can also be owned as property. But it is ontologically separate from human existence.

The assumptions that inform this paradigm have no parallel in Kamea thought. The people with whom I worked do not rely on physiological reproduction to track social relationships through time. Instead, intergenerational relations depend on the links that one forms with the nonhuman environment. In the absence of an ideology based on descent, the organic world serves as a point of attachment between human generations, and this gives Kamea sociality its decidedly patrilateral cast.

By means of conclusion, it is interesting to reflect upon a recent set of exchanges that have taken place within the discipline of biology. In *The Origin of Species* Darwin wrote about a master phylogenetic tree on which all life forms (both living and extinct) could be plotted on a single genealogy. However, it now looks as though the "kingdoms" of nature have been exchanging genetic material for a very long time (Pálsson 2005; cf. Doolittle 1999; Pennisi 2003; Syvanen 2002), thereby introducing taxonomic chaos into orderly traditional biological classification. For example, there is evidence that such "horizontal" gene transfers played an important role in the emergence of the earliest kingdoms of life a few billion years ago and have continued to shape the form of bacterial genomes over the past 100 million years (Syvanen 2002: 380). Similarly, according to researchers at the Human Genome Project, many human genes seem to have been acquired laterally from bacteria. According to some, such horizontal transfers of genes (resembling the much feared transgenic species crossings of the contemporary era) may be the major evolutionary source of true innovation (Pálsson 2005: 24).

Lateral gene transfer gives us an additional perspective from which to reflect upon Kamea sociality. First, it prompts us to question the idea of a universal evolutionary tree in which relatedness is established vertically and exclusively on the basis of descent. Second, it provides us with a point of departure from which to think about the notion of the "family tree" itself.

In the imagination of Europeans and North Americans, the "genealogical tree" serves as a representational device; it is a convenient way to map a different and more fundamental (i.e., essential) level of reality. Family trees are symbols. That is all they ever can be given that the world as perceived by Western audiences dissolves into a set of innate distinctions. Human reproduction has its own unique developmental trajectory that is markedly different from that followed by other life forms. Kamea provide us with a glimpse into a world in which the "family tree" takes on literal, rather than figurative, meanings.

TWO

Insubstantial Identities

In the summer of 2002, newspapers throughout Europe and North America featured the story of a young white woman—known only as Mrs. A—who had given birth to black twins in the United Kingdom. Mrs. A and her husband had been undergoing a treatment for infertility known as intracytoplasmic sperm injection at the Assisted Conception Unit of Leeds General Infirmary (L. Taylor 2002; M. Taylor 2002). The procedure involves harvesting eggs from the mother, fertilizing them outside the womb with live sperm taken from the father, and implanting the resulting embryos in the womb of the intended mother-to-be. Despite extensive protocols designed to prevent such a possibility, a "mix-up" occurred, and sperm taken from a black man (Mr. B) was used to fertilize the eggs of Mrs. A (McKillop 2002). Dame Elizabeth Butler-Sloss, president of the family division of the high court, said that this meant that, while the white woman was the children's biological mother, the black man (and not her husband) was their father (Firth 2002). "After the birth, it was noticed that the color of the children's skin was different from that of Mr. and Mrs. A: they were obviously concerned and made inquiries," said the judge (quoted in Seamark 2002). Subsequent genetic testing revealed that Mr. A was not the twins' biological father.

The circumstances surrounding the birth of the "black" twins generated a great deal of public interest, both within Britain and overseas. Described in the press as an "unmitigated disaster" (Winston 2002), a "tragedy" (Anonymous 2003), a "nightmare" (Birkett 2002), and a "circus of human sorrows" (Birkett 2002), the incident cast a shadow over the booming fertility industry.[1] It also established a new and important precedent in English law concerning the legal definition of parents.

According to the British Human Fertilization and Embryology Act passed in 1990, any woman who carries and delivers a child is considered to be the "mother" of that child, irrespective of genes (Freeman 2002). Thus, in cases of gestational surrogacy, for example, it is the woman who gives birth, and not the woman who donates the egg, who is defined for legal purposes as being the mother of the child (Cannell 1990: 672). The woman who provides the egg may attain legal parentage only by adopting the child. The definition of fatherhood is equally specific in this act. "In the case of a married couple . . . the husband in that relationship will always be considered the legal father of a child conceived through fertility treatment using someone else's sperm," unless he was legally separated from his wife at the time of the treatment, or if he can prove that he did not consent to the treatment in question (Dearle 2000; cf. Ripston 1990). Neither of these conditions applied to Mr. A.

In rendering her decision, which became public in February 2003, Dame Butler-Sloss ruled that the black man whose sperm was mistakenly used to produce the babies was to be treated as the legal father of the children (Anonymous 2003). She added, however, that there was no question of the twins being removed from the white couple. "This is a tragic human story of two families trying to come to terms with the consequences of the mistake. Although [the black man] was clearly not a consenting sperm provider . . . it is, on the facts of this sad case, only the use of his sperm that connects him with the twins" (Butler-Sloss, quoted in Anonymous 2003). The judge went on to say that Mr. and Mrs. A could formalize their position as legal guardians by having Mr. A adopt the children. At the time of this writing, it was not yet known if Mr. B intends to apply for parental rights or whether he will oppose the adoption.

The implications of the aforementioned case are enlightening to consider from the perspective of contemporary debates on Euro-American kinship. Since Lesley Brown gave birth in 1978 to her daughter Louise—the world's first "test-tube" baby—the expectation has been that the expanding use of new reproductive technologies (NRTs) would lead to a radical shift in terms of how people in Europe and North America conceptualize the link between parents and offspring. Dwight Garner—an editor for the periodical *Graphistock*—asserts, for example, that the development of new technologies is making it "harder than ever" to define the meaning of parenthood. He writes: "Where does DNA end and where do attachment and obligation begin? In an era of surrogacy, adoption, artificial insemination, in-vitro fertilization and genetic testing, as the lines on many of our family trees prepare to be smudged or redrawn, biology is less and less decisive" (Garner 2001: 20). Similarly, in a 1992 paper published in *Current Anthropology* Chris Shore states: "What makes these technologies so controversial [are] their social and legal implications. Not only do they 'crystallize issues at the heart of contemporary social and political struggles over sexuality, reproduction, gender relations and the family' (Stanworth 1987: 4) . . . but they challenge our most established ideas about motherhood, paternity, biological inheritance, the integrity of the family, and the 'naturalness' of birth itself. This may be one reason why they appear threatening to so many" (Shore 1992: 295).

The case involving the black twins suggests precisely the *opposite* conclusion. Far from casting Euro-Americans adrift from their longstanding adherence to biology as the underlying basis of kin connections, Dame Butler-Sloss's decision that Mr. B was to be recognized as the legal father of the children suggests a *deepening* of substance-based connections as the defining criteria of motherhood and fatherhood.[2]

Although the twins' case was the first of its kind in Britain, it is not without precedence. Nor has it been the last to make headline news. Significantly, the way in which these other cases have been resolved in the courts similarly points to a growing fetishization of substance as the perceived essence of kinship:[3]

· In 1993, Wilma and Willem Stuart became the parents of twins—one black and one white—after undergoing in-vitro fertilization (IVF) at the University Hospital in Utrecht. A year later, DNA tests revealed that the hospital had mistakenly mixed the sperm from the woman's white husband with that of a black man from the Dutch Antilles who had been undergoing treatment at the same clinic (Lyall 2002). When asked about her reaction to the "mix-up," Wilma Stuart said: "We don't oppose IVF. But we have just added a black line to our family. And we will still have a black part of our family in 100 years" (quoted in Anonymous 1995). The Stuarts were permitted by the courts to keep both children.

- In 1998, Donna Fassano gave birth to twins—one black and one white—after undergoing in-vitro fertilization treatment at a New York clinic. An investigation into the incident revealed that Fassano had been implanted with two embryos: one created by her egg and her husband's sperm, and the other by an African-American couple who had been seeing the same specialist at the same time. When genetic tests revealed that the black couple were the biological parents of the child, Fassano agreed to "*give the child back*," stating that she "did not want to separate a child from his natural parents" (Anonymous 1999; Wulf 1999; italics added).

- In the fall of 2002, in the wake of the black twins case described earlier, a London hospital became the source of a media frenzy when it was revealed that two women had been implanted with the "wrong" embryos while undergoing IVF procedures (Marsh 2002). Because identification labels were not properly checked, one patient's healthiest embryos were implanted in a second woman, whose embryos, in turn, went to a third woman.[4] When the mistake was detected, the two women were ordered back to the hospital where an emergency technique was carried out to "flush" the embryos from their wombs. The women were also given drugs to ensure that there was no risk of pregnancy (Allen 2002; Pook and Martin 2002).

These cases are highly suggestive. As Marilyn Strathern has argued, the present epoch has witnessed an exaggerated emphasis on the biological idiom. Until recently, she says, "the naturalness of the procreative act was not sufficient to establish real relations. There was also the issue, we might say, of the naturalness of social status. Reproducing one's own did not literally mean one's genetic material: one's own flesh and blood were family members and offspring legitimated through lawful marriage" (M. Strathern 1992a: 52). Strathern cites Northcote W. Thomas, who observed in 1906 that under English law, the father of an illegitimate child was not "kin" to it, despite the blood tie that existed between them (cf. Schneider 1984). Thus, "it was improper for the offspring of an illicit relationship to go into public mourning for their parent. The fact of their grief was irrelevant: to claim kinship was a public (social) act" (M. Strathern 1992a: 52). This perspective can be productively compared to the emerging ethos of the twenty-first century: today, children are "given back" or are flushed from the wombs of unsuspecting women if it can be established that they lack a substance-based connection to the woman who gave (or, is about to give) birth to them.

The revalorization of biology that has accompanied the expanding use of NRTs carries with it a number of important implications. As an orienting device, biology relies on a set of interrelated assumptions concerning the nature of temporality, what connects (or disconnects) people as "relatives," and what gets (re)produced through time under the guise of "kinship." It also privileges certain kinds of relationships (i.e., parentage) as being the "most important" in a generative sense, while downplaying others (including, for example, siblingship) as comparatively insignificant. These ideas are gaining a new kind of purchase in the twenty-first century and are coming to be expressed in Euro-American kinship configurations in unprecedented ways.

By means of illustration, I turn to two emerging uses of NRTs: posthumous reproduction and parent-child gamete donation. Two decades ago, these practices would have been the stuff of science fiction; today, they are common requests at fertility clinics. What makes these practices particularly interesting is their capacity to reveal many of the conceptual side effects that come into play by viewing kinship as an embodied relation.

In 1997, Julie Garber, a California real-estate developer, became the first woman in history to receive the dubious honor of nearly becoming a mother from beyond the grave. Three years earlier, the twenty-six-year-old woman learned that she was suffering from acute lymphoblastic leukemia. Before embarking on a course of chemotherapy that would render her infertile, she arranged with a sperm bank to have a dozen of her eggs fertilized and the resulting embryos frozen (Pepper 1996; Peres 1997a, 1997b; Weiss 1998a, 1998b). Her hope was to have the embryos implanted in her uterus after recovery. When Garber died, her parents hired a surrogate mother to bring their daughter's ungestated offspring to term, an act they said fulfilled one of Julie's last wishes. Julie's parents had agreed to give any resulting offspring to one of their remaining children to raise. After three unsuccessful attempts, however, their quest came to an end in December 1997 "when the last of Julie's embryos was rejected by the surrogate mother's body several weeks into the pregnancy" (Siegelitzkovich 1998).

Although the Garbers' decision to try to bring their daughter's child to term raised serious ethical questions, the practice of harvesting gametes from the recently deceased has become fairly commonplace in Europe and North America.[5] A 1997 survey found that fourteen clinics in the United States had honored requests for sperm to be collected from the deceased (Andrews 1999b: 226). Cappy Rothman, a leading advocate in the use of this technology, defended the practice before a 1997 meeting of the New York legislature, where the question under debate was: should the sperm of a man be used without his expressed prior consent? Rothman argued that there is less grief for the wife and family members of the deceased if the sperm is saved (Andrews 1999b: 227). He went on to tell legislators, "In one case where a man died by gunshot and I collected his sperm, his family followed me to the sperm bank and were consoled by seeing his motile sperm under the microscope. To console families in that way at a time of grief and tragedy is clearly part of my responsibility as a healer" (quoted in Andrews 1999b: 227).

The practice of retrieving sperm from the dead has become so common in the United States that the American Society of Reproductive Medicine has developed a protocol, "Posthumous Reproduction," for dealing with it (Andrews 1999b: 226).[6] Requests to harvest the gametes of the recently deceased are not made simply by mothers and fathers of the deceased (as in the case of Julie Garber), but also by spouses, siblings, boyfriends, and girlfriends (see Andrews 1999a and 1999b for further details). In 2002, the British government gave its backing to a bill that will allow women who have children using the frozen sperm of their dead husbands to register the latter as the legal fathers of any children they may bear by this method (Brown 2002). The legislation was a victory for Diane Blood, thirty-seven years old, who fought a three-year battle in the United Kingdom for the right to have children using sperm from her deceased husband, but who was prevented under existing law from naming him as the father on the birth certificate (Brown 2002). Later that same year, Gaby Vernoff gave birth to a baby girl in a Los Angeles hospital. The child had been conceived using sperm that was taken from her husband Bruce thirty hours after he died from an allergic reaction to prescription medicine (Home 1999). Since that time, there have been several children born to men who had no idea prior to their death that fathering a child was even a remote possibility in their future (Andrews 1999a, 1999b: 222–36; Cohen and Day 2000; Ward 2002).

The practice of posthumous reproduction highlights several important assumptions that exist at the heart of Euro-American kinship configurations: namely, the idea that individuals (re)produce individuals. For Europeans and North Americans, contained within the genetic substance of all living human beings is a microscopic image or blueprint of that person in miniature. By engaging in further acts of procreation, some part of the original person is believed to "live on," embodied in a new individual. Howard and Jean Garber pursued elaborate means (not to mention great financial expense) in order to preserve something of their daughter Julie. That their attempt ultimately failed does not belie their original intention—they wanted to "hold on" to their daughter in the face of death.[7] Similarly, when women like Diane Blood request to be impregnated using the sperm of their late husbands, they are not simply motivated by a desire to become a "mother." Were this the case, their aims could have been achieved far more simply by either adopting a child or by undergoing artificial insemination by anonymous donor. The reason Euro-Americans do not want to let the gametes of a deceased loved one go is because Westerners imagine that the dead will achieve a kind of immortality through his or her children.[8]

Exactly how is such a remarkable feat understood to be possible? Why does it seem "natural" to Europeans and North Americans that individual identity gets carried forward in time through further acts of reproduction? The practice of parent-child gamete donation illustrates some of the mechanisms that are understood to be involved.

In November 2000, a Sunday newspaper in Britain reported that women were being impregnated with sperm that had been donated by the father of their infertile partner. This practice, which has been carried out in Japan for many years (Anonymous 2000; Takagi 2000), has recently become available as an option to infertile couples in Europe and North America. Journalists Jason Burke and Paul Harris (2000) quote a "senior medical authority" as saying that the practice is "logical, appropriate and ethical." They also cite Anne Widdencombe, the shadow home secretary, who warns: "The tangled webs that we are weaving for future generations are just horrifying. This is not something I would ever endorse" (quoted in Edwards forthcoming). Widdencombe's report goes on to detail the perceived repercussions of this practice: "This treatment will change every single relationship within the family" (Burke and Harris 2000). More specifically: "Children conceived through such a process would effectively become the siblings of their father, who would be expected to raise them. Their biological father would also be their biological grandfather" (Takagi 2000). Conceptualized on the one hand as "logical and appropriate" and on the other hand as "horrifying," these dissenting opinions reveal some important assumptions concerning how consanguinity is imagined by English-speaking Euro-Americans (Edwards forthcoming).

In pursuing this theme, I draw upon Jeanette Edwards's recent work in Alltown, a small city in the north of England. Since 1992, Edwards has been interviewing Alltown residents concerning their reactions to emerging genetic technologies. Following in the wake of the Burke and Harris article, she asked men and women to express their views concerning parent-child gamete donation. Many of her interviewees readily assented that this practice evinced "a certain kind of logic." According to one man, "They've got a point. It would be the nearest thing genetically." "Of course," he added, "with a brother [it] would be even more likely to get the same type." Similarly, one woman with whom Edwards spoke admitted that she could understand "that blood thing." She went on, "For a man it would be better to receive sperm from a close relative than from a stranger out of a bottle . . . because then you'd know" (Edwards forthcoming). These comments highlight the importance of a substance-based connection in terms of how Euro-Americans conceptualize what it means to be "related." As Edwards points out:

> Being able to trace one's genetic connection is perceived to be good in itself; alleviating insecurity and diminishing unpredictability. It is what is perceived to be shared in terms of genetic "make-up" that makes a father a suitable sperm donor, and a brother even more suitable. But it is also in the recognition facilitated by what will be seen as family resemblance. The child born from the gametes of kin rather than an anonymous donor will be of the "same type" which will be manifest itself in behavior and appearance that is familiar. (Edwards forthcoming)

Sidestepping one link (i.e., the intending father) in terms of genealogical connectedness appears to matter little from this perspective. All members of a given family will to share the "same substance" in common when it is traced as a continuous line of descent. Indeed, the goal of keeping substance-based connections intact, rather than losing them, is what motivates intrafamily gamete donation in the first place. In describing this practice to the press, Dr. Atsushi Tanaka, who carries out this procedure in his own fertility clinic writes: "If a child ever wants to know who their 'real' father is, the easiest way to handle it would be to tell them that they were born using their grandfather's sperm and that their father's blood is flow-ing through their veins" (Takagi 2000). Substance, in short, imparts an imagined essence that links all persons who share it as embodied kin.

If parent-child gamete donation is seen on the one hand as "logical and appropriate," it is not without its critics. One of the most common objections raised about this practice is that its use "confounds" naturally occurring degrees of relatedness. If kinship is substance, and if substance is something that gets transmitted through time, it should be possible to specify the "closeness" and "distance" of kin with something approaching mathematical precision. Marilyn Strathern has described how this logic works in terms of Euro-American conceptions of kinship:

> The flow of blood is at once like a moving stream (and cannot travel backwards) and like a substance that can be infinitely divided into parts. Hence the "dilution" of any one stream that comes from mixing, rendering any one individual an amalgam of blood. The procedures for working out their proportion are simple; one divides one's blood in half by parents, into quarters by grandparents, eights by great grandparents (cf. Wolfram 1987: 13). Born of a mother with two English parents and of a father with a Welsh and English parent makes me one quarter Welsh. But the actual "flows" have been rendered invisible—one sees instead the traits each individual displays. Indeed, in popular belief, the parts that an individual person "gets" from either mother or father may be thought of as parts of other ancestors that "show" in descending generations. (M. Strathern 1992a: 80)

Donations of sperm and eggs between parents and children challenge the treelike progression through which reproductive events are understood to unfold in the West. In the minds of many, a "confusion" of kinship identities is likely to follow in turn. Dr. Samantha Gothard, a psychiatrist at London's St. Anne's Hospital, expresses her concern in the following way: "[The practice] makes a father's son his biological half-brother and a child's biological father his or her grandfather" (quoted in Burke and Harris 2000). Similarly, as one Alltown resident cautions: "Emotionally—the granddad knows that that's his child—be it male or female—is he strong enough to just sit back? He's got to think a lot of [his son]. He's got to be very parental, you know, to do that in the first place . . . so, can he just take a back seat and let his son and daughter-in-law bring this child up without any sort of interference? I don't know" (quoted in Edwards forthcoming). In the dominant Euro-American paradigm, social roles and behavior emanate "naturally" from biological states. Indeed, biology does not just define an abstract social relation, it is often argued to promote certain kinds of behaviors between persons related as "kin" (cf. McKinnon and Silverman 2005: 5; Nelkin and Lindee 1995: 58–78). Given this, would a "brother" really be able to care for a child in the same way that a "father" might? Would it be possible for a "grandfather" to step aside and let his "son" raise a child that he, himself, begat? Parent-child gamete donation is seen to threaten "naturally occurring" social behaviors.

Even more disturbing to some is the possibility of reversing the direction of donation: of having children provide gametes to their parents for procreative purposes. Susan McCall, a youth and community worker in Alltown, tries to imagine what it would mean for a daughter to donate ova to her mother:

> It is almost like going back in time. I'm not sitting here saying that they'd be wrong to do it. . . . If a daughter donates an egg to her mother, there is no reason why there is going to be a genetic problem with that, is there? Because she is purely a carrier. I'll tell you what, if it was written on a piece of paper, I'd say, "Oh yes, that's fine," but actually thinking through the relationships is tricky (pause). It is almost like saying the father is fertilizing the daughter's egg (pause), which does bring up all sorts of fears about inbreeding—even if it isn't actually incest. (Quoted in Edwards forthcoming)

McCall draws our attention to one additional concept that underlies Euro-American kinship configurations: directionality. Substance should flow in one direction and one direction only. Just as care and nurture move from parents to children, so too does the flow of genetic material. To reverse that flow—to send substance back in the direction from which it came—elicits feelings of intense unease. It conjures up images of incest, perhaps the most loaded conceptual construct in Euro-American kinship ideology.[9]

By means of summary, this brief foray into the world of assisted conception technologies (ACTs) has helped to reveal several key assumptions upon which dominant Euro-American notions of relatedness rest. (1) Kinship is equated with biology; this holds true today perhaps even more than it did in the past. (2) Individuals (re)produce individuals. (As we shall see, this contrasts markedly with Melanesian conceptions, where the preeminent focus is on the reproduction of relationships.) (3) The parent-child tie is the most important in a generative sense: it is this connection that carries sociality forward. (4) Generations are tied to one another by an embodied link.[10] (5) Kinship entails a particular kind of temporal sequence. Relationships unfold along a unidirectional time line and the succession of generations cannot (or at least, should not) be reversed or recycled.

This model has structured more than our understanding of relatedness: it has figured centrally in the development of anthropological theories of sociality and has often negatively influenced our ability to grasp the contours of alternative social realities.

ALTERNATIVE CONCEPTIONS

Kamea conceptions of kinship offer a compelling foil to the cultural logic of Euro-American representations. If the West has never been able to dispense with the idea that consanguinity is, at its core, a biological phenomenon (Schneider 1968, 1984), Kamea seem prone to move in the opposite direction by virtue of their tendency to divorce sociality from the body and its capacity to procreate.

In the last chapter, we saw that Kamea do not frame intergenerational relations as a product of genealogical connections. Lineal continuity is not imagined as a history of procreative events. Rather, a boy creates his own place in the nexus of remembered social relationships by virtue of the ties that he forms with the land and a variety of nonhuman resources. Male sociality acquires permanence and historical significance only by being objectified in and through the nonhuman environment.

In the following pages, I take this argument one step further. By examining indigenous ideas surrounding the reproduction of human beings, we shall see that the entire notion of "descent" as a process involving the transmission of bodily substance has little significance in terms of how Kamea conceptualize and act within the universe of social relationships in which they participate. If Euro-Americans see a close connection between the building of bodies and the building of social ties, Kamea see instead two distinct processes.

To claim, as I am doing, that Kamea do not draw upon a biological framework as the basis of kin connections will hardly seem like a novel idea. It has, after all, been more than thirty years since David Schneider (1968, 1984) first shocked the anthropological world with his seemingly heretical statement that kinship, as we knew it, did not exist. Instead of analyzing a universally existing phenomenon, anthropologists had been imposing their own views of social life elsewhere. Furthermore, recent years have witnessed an efflorescence of works that question, as Schneider did, the universal importance of physical reproduction and the bonds it creates among human beings. Several of these newer works attempt to document the so-called "processual" nature of kinship by showing how relatedness is a gradually acquired state that can be built through time by nonsexual means.

Janet Carsten's (1995, 1997, 2001) work with the Malays of Southeast Asia is representative of this growing trend. In several recent publications, she argues that people on the island of Langkawi become "kin" through living and consuming together in houses. "Bodily substance," she writes, "is not something with which Malays are simply born and remains forever unchanged. . . . [Instead] it gradually accrues and changes throughout life as a person participates in social relationships" (1995: 225). Mary Weismantel adopts a similar stance in her analysis of adoption among the Zumbagua of highland Ecuador. In a 1995 paper, she writes: "The physical acts of intercourse, pregnancy, and birth can establish a strong bond between two adults and a child. But other adults, by taking a child into their family and nurturing its physical needs through the same substances as those eaten by the rest of the social group can make that child a son or daughter who is physically as well as jurally their own" (Weismantel 1995: 695). Especially critical in this process of "making kin" are intentional acts of feeding and sharing:

> Flesh is made from food, and especially from different grains and tubers, each of which has its own characteristic effect on the human body. Eating cooked grains raised by a household on its own land and harvested and processed through family labor results in a body and a self that have been shaped by work and skill invested in the farm. When men bring home foods bought with wages, these foods change the bodies of family members too. . . . Those who eat together in the same household share the same flesh in quite a literal sense: they are made of the same stuff. It is when young Iza's boy has eaten so many meals with the family and his while body is made of the same flesh as theirs that the bond will be unquestioned and real to the boy and his [adopted] family. (Weismantel 1995: 695)

Proponents of the processual model dedicate a great deal of attention to postnatal means of creating and validating bonds of corporeal consubstantiality. Kinship ties that Euro-Americans typically understand to be biological in nature (such as parenthood and siblingship) are argued to be "made" rather than a product of birth.

Despite the seeming novelty of these newer theoretical formulations, they continue to rest upon two underlying ideas: first, that kinship is a bond of substance, and second that it unites two or more people in a "physical" relation.[11]

In what follows, I challenge Euro-American assumptions about substance as being central to relatedness. I address this theme by asking the following question: need kinship always be conceptualized as a "material" bond (whether created or given at birth) between two people? Put somewhat differently, is there a way of thinking about consanguinity that does not ground it in bodily connectedness, or at least that imagines contexts where it is not embodied as a substantial link between people?

In the remainder of this chapter, I consider this question from the perspective of Kamea. We shall see that Kamea sociality is founded upon a distinction between what might be called "substantial" and "insubstantial" identities.[12] Here, bodily substance occupies an important place in terms of how *same-generation* (i.e., sibling) relations are conceived, but is of little or no importance to how they are tracked through time. It follows that Kamea sociality bears little resemblance to Euro-American kinship configurations. Because Kamea women (specifically mothers) are the source of the sibling tie, this chapter considers in more detail what I call female sociality. It is, thus, intended to complement the preceding discussion of male sociality, by showing how the two (and distinct) gendered modes of relating combine to shape Kamea social life.

In pursuing these themes, my discussion will unfold along the following lines. After briefly considering the significance of sibling studies in Melanesian ethnography, I turn to an examination of Kamea ideas surrounding procreation. I argue that Kamea draw upon an image of "containment" (i.e., being enclosed within a particular womb) as the key image through which the mother-child bond is conceived. Through an examination of the diverse contexts within which this image becomes relevant (including marriage and childprice payments, kinship terminology and myth), I demonstrate that women are associated with the elicitation of lateral (or intragenerational) social relations.

The final section of this chapter turns to a consideration of how male and female modes of sociality intersect. Through their productive combination, an analogic equivalence is postulated between siblingship and affinity that makes the analysis of one invariably the analysis of the other (cf. J. Shapiro 1985). It is the ongoing transformation of one into the other that gives Kamea social life its forward-going momentum. If a genealogical model privileges the reproduction of individual persons, Kamea social life is focused, instead, on the ongoing transformation of different sets of relations. As we shall see, several important repercussions follow logically in turn.

ON THE NATURE OF THE SIBLING TIE

Anthropological analyses of siblingship have typically begun, as have studies of kinship more generally, with the notion that the conjugal tie (the assumed nexus of biological reproduction) forms the basis from which other types of social relationships flow. Indeed, Radcliffe-Brown's early formulation of the "principle of the unity of the sibling group" was based on the idea that siblings constitute a homogenous whole by virtue of their common parentage, and hence they are in an identical position in the genealogical grid. In Radcliffe-Brown's words, this principle "refers not to the internal unity of the group as shown in the relations of its members to one another, but to the fact that the group may constitute a unity for a person outside it and connected with it by a specific relation to one of its members" (Radcliffe-Brown 1950: 23–24). Siblings were seen to be interchangeable parts of a sociological whole because their shared descent from the same parents placed them within an analogous place in the social structure.[13]

In the kinship systems studied by Radcliffe-Brown (predominantly those derived from African and Australian ethnography), the principles of descent and complementary filiation were taken to be the invariant building blocks on which the entire edifice of social structure rested. Unilineal descent furnished order and continuity through time by articulating an unambiguous rule for group recruitment (cf. Marshall 1981b). Individuals were assigned to clans and lineages on the basis of birth and these, in turn, structured marriage arrangements and property relations.

As I have already indicated, this analytic framework had little explanatory power when it came to the social systems of Melanesia. Here, descent failed miserably to predict the composition of local groups and siblingship seemed to carry more than its fair share of the social and symbolic load. Indeed, in the minds of many early ethnographers, Melanesian societies represented a nearly perfect inversion of the principles originally set out by Radcliffe-Brown. Instead of having siblingship derive from rules of descent, the reverse was often argued to be the case. In a now classic formulation of this interpretative framework, Kenelm Burridge writes: "In Tangu society, it is those values attached to siblings and siblingship which seems to reveal what is relatively constant or fixed. *It would be true to say in Tangu today that a person's descent is of small significance to him but that relationships with sib-lings are of vital importance.* Briefly, that siblingship is the determinant that descent might have been expected to be" (Burridge 1959: 128; italics added). Burridge goes on to argue that "descent was probably always calculated from siblingship" in Tangu, and that marriage rules take siblingship rather than descent as their point of departure.

Since Burridge's seminal article first appeared, many field-workers have supported the view that sibling ties are of particular relevance in Pacific Island societies. In one of the most detailed studies of siblingship in Oceania, Ray Kelly (1977) demonstrated that among the Etoro of the Southern High-lands, Papua New Guinea, sibling relations supercede those based on descent as the primary organizing principle of social life (Marshall 1981a). This premise is echoed in Schieffelin's study of the neighboring Kaluli: "Kaluli ties of sibling relationship are in contradiction to those traced by descent (by genealogical reckoning) . . . and the sibling relationship takes precedence over descent when ever the principles are in conflict" (Schieffelin 1976: 56). Wagner (1967) also comments on the significance of the sibling bond in his description of Daribi clanship, particularly, in terms of the rights that a man is recognized as holding vis-à-vis the offspring of this sister.

Kamea represent a particularly interesting case to consider in terms of the foregoing discussion. For them, genealogical reckoning does not furnish a means of assigning people to groups; indeed, as we have already seen, there are no independently existing groups to which people may be recruited. To grasp how Kamea think about siblingship and its relationship to how social ties are tracked through time, we need to consider indigenous ideas sur-rounding birth and conceptions of relatedness. These themes are taken up in the next section.

CONTAINMENT AND THE "ONE-BLOOD" RELATIONSHIP

A Westerner approaching Kamea sociality for the first time is likely to be lulled into a false sense of security fostered by the ease with which indige-nous formulations seem to translate into our own. Like most Papua New Guineans, Kamea believe that repeated acts of sexual intercourse are nec-essary to create a child. Conception is said to take place when sufficient quantities of a man's reproductive fluids (*iya coka*) both mix with and are encompassed by the fluids of his wife (*panga coka*). This outer female cov-ering will eventually form the skin and surface blood vessels of the child while the father's semen contributes to the making of bones and internal organs. Yet, although both parents contribute substance to the making of a child, this is not seized upon as a compelling feature of the parent-child relationship. Unlike Europeans and North Americans, Kamea do not equate the building of bodies with the building of social relations.

This difference between Western and Kamea formulations is strikingly revealed in the fact that although Kamea use the idea of "one-blood" groupings to differentiate persons in their social universe, this expression does not refer to intergenerational connectedness; instead, it speaks to the shared experience of having been nurtured within the same woman's womb. The Kamea term *hinya avaka* may be glossed as "one-bloodedness" and is used to refer to a sibling set. Other than using this term, there is no way for a speaker to refer to all of his or her siblings as a single, undifferentiated category. All other terms in the Kamea language use sex and age to distinguish between different types of siblings. Yet, despite their use of "blood" as an idiom of relatedness, Kamea formulations of kinship bear little resemblance to those of the West.

Any children that a woman bears, regardless of who the father might be, are said by Kamea to be "one-blood" with one another. The same, it should be pointed out, does not hold true of a man. Should a man have more than one wife during the course of his life, any children that he has with these separate women will not be spoken of in terms of the "one-blood" relationship.[14] Only persons born of the same womb are *hinya avaka*. To be "one-blood," then, is to have originated from the same maternal container.

This same notion of "one-blood" groupings effectively *separates* rather than connects a woman to her offspring. I initially confused the notion of "one-bloodedness" with Western ideas concerning the inheritance of biogenetic substance and assumed that the expression referred to the cultural fact that blood is the female contribution to conception. Kamea, as it turned out, did not share my fascination with the field of biology. Neither a woman nor a man is considered to be "one-blood" with their children; the term refers exclusively to having issued from the same woman's womb.[15] Thus, one's mother, for instance, would be "one-blood" with her own "true" (tru) siblings, but not with any person in the ascending or descending generation. Women bestow upon their children a horizontal type of relatedness that is imagined in terms of "containment," rather than the lineal transmission of bodily substance.

In the chapters that follow, I document how this imagery of containment is reflected in a wide variety of social practices. The Kamea men's cult, for example, is focused on the lengthy and arduous process of detaching a boy

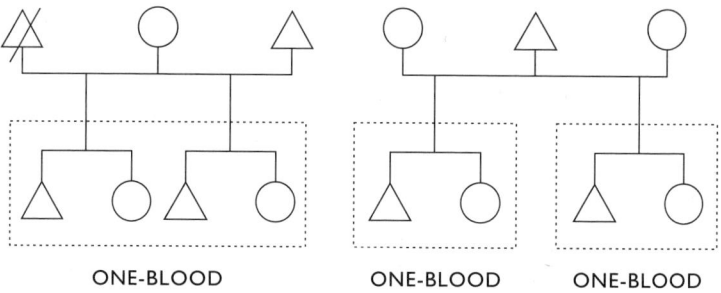

ONE-BLOOD   ONE-BLOOD   ONE-BLOOD

Figure 2. The "one-blood" relationship. Males are represented by trian-
gles and females by circles. The left-hand side of the diagram represents
two sets of children with different fathers (the first one deceased, represented by the canceled triangle) but the same mother, and who are
therefore considered "one-blood." The right-hand side represents two
sets of children with the same father but different mothers; only the
children with the same mother are thought of as "one-blood."

from the containing influence of his mother, a state that is perceived to exist
long after birth. Indeed, so close is the connection between a boy and his
mother that the eating habits of one are seen to directly affect the health and
well-being of the other (Bamford 1998a, 2004, 2006). Initiation brings
about a process of "decontainment," thereby allowing a boy to act as an autonomous male agent. In the cultural logic of Kamea, to be female is associated with the capacity to "contain," while to be male is, among other
things, to exist as a "decontained" social being.

This system of ideas has a number of important implications. For while
Kamea see human life to have its origin in sexual reproduction, substance-
based idioms are not used to frame human relationships through time. Although the people with whom I worked had no difficulty in specifying who
their mother and father were, in the sense that these persons contributed
the necessary elements toward conception, the parent-child tie is not imagined as corporeal. Substance, in short, lacks the temporal dimension that
makes genealogy coterminous with reproduction in the West. As we have
already seen, Kamea *do* have a means of tracing social relations through
time, but this is not based on genealogical connections; instead, it is based
on human-environmental relations.

From the foregoing, it is evident that Kamea men and women have
different modes of relating that can be glossed in terms of a contrast between "lineal" and "lateral" sociality. For men, the movement through the

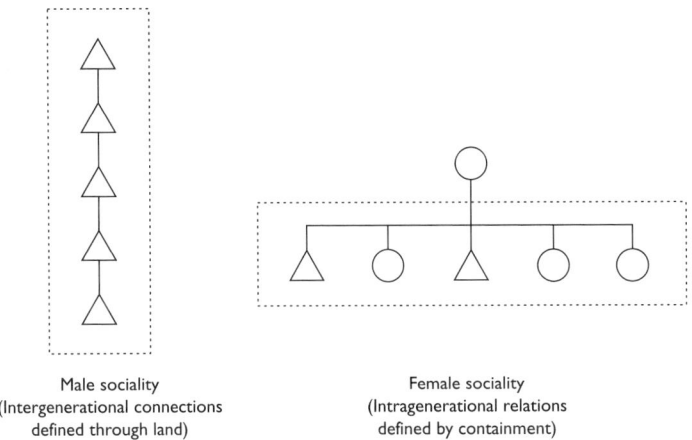

Male sociality
(Intergenerational connections
defined through land)

Female sociality
(Intragenerational relations
defined by containment)

Figure 3. Male and female relational configurations. Men are associated with vertical, or intergenerational, ties; women are associated with horizontal, or intragenerational, ties.

life cycle is interpreted as a movement toward planting: toward constructing intergenerational links by acting directly on the land. Female sociality follows an entirely different track. Women are not connected to their offspring vis-à-vis lineal connections; rather, the mother-child relationship is imagined in terms of an opposition between the "container" and the "contained." In terms of the *kinds* of relationships that each sex is perceived to be capable of eliciting, women are associated with the creation of horizontal (i.e., intragenerational) ties and men with vertical (i.e., intergenerational) ones.

Defined as such, an interesting problem presents itself. Kamea social life is based on two gendered modes of relating, but neither is capable of its own reproduction. Although lineal continuity is "productive" in the sense that it possesses a temporal dimension, taken on its own it cannot re-create the conditions for its own existence. New persons need to be invested in the land, and these cannot be produced through the exclusivity of same-sex links. Relationships defined through women are subject to similar limitations. Mothers, in Kamea thought, give birth to sets of relationships—to "one-blood" similitude—rather than to sons and daughters per se. Neither male nor female modes of relating are self-sufficient in terms of their regenerative capacity.

How cross-sex siblings come to be sent on these differentiated trajectories—that is, how they come to be seen as gendered beings with distinct reproductive potentials—is in no small measure a product of affinal relations. It is consequently to an examination of the marriage bond that I now turn my attention.

GENDERED MODES OF RELATING: AFFINITY AND THE ELICITATION OF CONTAINMENT

Thus far, my discussion of Kamea sociality in this chapter has focused primarily on the sibling relation. It should be emphasized, however, that siblingship is not an isolable phenomenon; it is intimately tied to indigenous understandings of gender and parenthood (Yanagisako and Collier 1987). In this section, I consider the relationship between siblingship and affinity. I shall argue that these two types of relationships cannot be rigidly separated from one another and that Kamea sociality rests not on the production of persons, but rather on the temporal transformation of one mode of relating into the other.

In his work with the people of the northern Gilbert Islands, Berndt Lambert (1983) argued that brother/sister and husband/wife represent two alternative ways of ordering male-female relations in the Pacific. He writes: "The complementarity of brother and sister and the cooperation between husband and wife represent contrasting forms of mutual dependence between men and women; both rest on conceptions of the activities proper to each sex" (Lambert 1983: 151). Similar arguments have been set forth by Riviere (1971), Damon (1983), and A. Weiner (1992) in their analysis of other Pacific Island societies. The point I wish to make in the following discussion is somewhat different. Marriage for Kamea is not about reordering the gender-reproductive axis: *instead, it is instrumental in creating it.*

Anthropologists working in the highlands of Papua New Guinea have long centered their discussions of marriage around the social and symbolic importance of matrilateral payments.[16] More than simply defining rights to a woman and her children, such payments are often crucial in delineating the boundaries of social groups and become an important venue through which intergenerational continuity is created and reproduced through time (M. Strathern 1987; Wagner 1967, 1977; J. Weiner 1982). For many highland peoples, the need to make such payments is informed by the idea that men and women contribute equally in a procreative sense to the formation of a child. Maternal and paternal contributions of blood and semen at conception relate a child *"equally but in different ways* to his mother and father" (J. Weiner 1982: 9; italics original).[17] Yet, notwithstanding the perceived bilaterality of reproductive connections, children are normatively recruited to the social group of their father.[18] As recorded in several ethnographic accounts, matrilateral payments offset the competing substance-based claims of maternal kin by compensating the wife's group for the bond of blood that they share with the wife and the child. As Wagner (n.d.) has eloquently argued:

> Among the Daribi people of the Karimui area, the transformation of female productivity—objectified in terms of mother's blood—into male continuity—is a lifelong undertaking. It is effected through the presentation of elements of male productivity to the relatives of a woman on her behalf or on the behalf of her children. . . . A Daribi social unit can be understood as a set of people with a common interest in redeeming and maintaining its male continuity through the transformation of female nurturance and productivity contributed by others. (20)

For many highland peoples, matrilateral payments do more than substitute wealth for persons. An important constituent of these presentations is often meat (Wagner 1967: chap. 6; J. Weiner 1982: 9), which is widely seen as an exterior and partible aspect of male procreative substance. If consumed in sufficient quantities, for example, pork is believed to augment a man's limited supply of semen, thereby safeguarding his capacity to act as reproductively competent agent in the future. In terms of the overall exchange, then, wife givers relinquish a female kinsperson and the right to affiliate her children, but receive in return a partible adjunct of male procreative substance through which they can reconstitute themselves as a viable and enduring group.

The people of Titamnga share many features in common with other highland groups. Intergenerational continuity exhibits a decidedly patrilateral cast, as has been commonly reported of other societies in the region. Furthermore, the reasons Kamea give for making these presentations is essentially the same. Matrilateral payments, I was told, safeguard the health of the child and prevent negative feelings from arising on the part of maternal kin (cf. A. Strathern 1971a: 458; Wagner 1967: chap. 6, 1977: 634). Although the men and women with whom I spoke held markedly differing opinions as to what might happen if such payments were neglected, there was general agreement that a child would fail to thrive if obligations to maternal kin were not met.

In these and other respects, Kamea resemble the majority of highland peoples as they have been described in the ethnographic literature. However, here the similarities end. Neither bridewealth nor childgrowth payments are tied to a process of group affiliation, for, as we have already seen, there are no independently existing groups to which a child may be recruited. It follows that there exists no common pool of clan wealth that a man can draw upon to oppose the claims of his wife's kin.[19] Perhaps even more significantly the transmission of bodily substance between a parent and a child is not a compelling feature of Kamea kinship configurations. Although a mother (like a father) contributes bodily substance to her child at conception, this does not result in an embodied connection. Matrilateral payments cannot be about "buying off" maternal substance, insomuch as the child is not seen to house such substance in his or her body in the first place. To grasp the significance of these presentations, it is necessary to examine Kamea social life in greater detail, including those meanings that underlie the "one-blood" relationship.

As noted previously, the term *hinya avaka* as it is used by Kamea is one of undetermined gender. Other than using this term, there is no way for a speaker to refer to all of his or her siblings as a single undifferentiated category; all other terms in the Kamea language use sex and age to distinguish between different kinds of siblings.

In its ungendered connotations, *hinya avaka* appropriately parallels the way in which children are understood to enter the world. For Kamea, an individual's sex at birth is seen to be the outcome of a struggle between maternal and paternal substances in the womb. If the mother's contribution is stronger, the child will be a girl; if the man's contribution is stronger, the child will be a boy (cf. Rohatynskyj 1990: 439). Yet, if birth sex defines certain potentialities, these potentialities must be *activated* if they are to come to fruition. During the first few years of life, Kamea boys and girls are understood to be identical kinds of beings; they are dressed in the same way and are allowed to engage in the same type of activities. Furthermore, parents will refer to children up until the age of five or so by the term *child* (*imia*), only later qualifying this form of address with the adjectives *male* (*imia oka*) and *female* (*imia apaka*). "One-blood" siblings, then, are defined in the first instance by their underlying similarity. It is only later that gendered attributes come to be associated with these undifferentiated forms.

It is the negotiation and eventual fulfillment of a marital bond that helps to create and sustain cross-sex sibling relationships. If throughout much of the highlands, bridewealth and childgrowth payments have to do with acquiring rights to a woman's fertility, for Kamea they are focused instead on creating that capacity in the first place. The difference between "male" and "female," or "brother" and "sister," are not taken to be innate, but have to be created against an ungendered backdrop of "one-bloodedness," which furnishes first and foremost an original field of "sameness." For women, this takes place at marriage; for men, at initiation, which is the subject of the next chapter. The types of identities that emerge from these processes are twofold: male is differentiated from female, and what was once contained is differentiated from the container that once encompassed it. It is through this dual process of differentiation that agency, the capacity to act within the world as a reproductive being, is created.

It will be useful to consider bridewealth in greater detail, as these payments constitute the first of a life-long series of obligations to matrilateral kin. A typical series of exchanges unfolds as follows. Shortly after an infant girl's birth ceremony, her parents will be approached by the family of a prospective groom interested in negotiating a marriage contract for their son. Since Kamea evince a marked preference for second-degree cross-cousin marriage, it will be the children of "one-blood" siblings who sit down to arrange a match between their respective offspring.[20] This means that first-degree cousins occupy the space between two "containment" cycles. Although not "one-blood" themselves, the children of cousins will marry and reestablish the conditions for lateral relatedness. Seen through time, there is a transformation of relationships such that parents produce siblings, who produce cousins, who produce siblings. Then the cycle begins anew.

Today, brideprice generally includes at least some items of Western manufacture, including cloth, store-bought food, blankets, steel knives, and the like. In the past, it consisted almost entirely of garden produce and game, the latter being the most important constituent. If the parents of the girl are agreeable to the proposed suit, the family of the groom will initiate a series of presentations on their son's behalf. A <u>bilum</u> (string bag) of sweet potatoes will arrive one day, a few weeks later a pouch of store-bought tobacco. These presentations will continue on an informal basis for several years, until eventually the boy will begin to assume the responsibility of making these payments on his own behalf.[21] Items of bridewealth are presented to the mother of the girl, who will redistribute them to other relatives as she sees fit. As one new husband explained: "I give these things because it was hard work [for my wife's mother] to care for my wife. As she was growing up, she shit and pissed all over her mother. It is hard work for the mother to remove the child's shit. That is why we give these things."

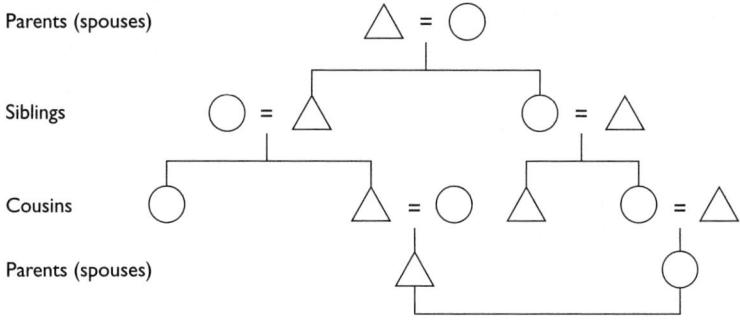

Figure 4. Kamea relational transformations. The unfolding of Kamea social life entails the reproduction of sets of relationships, rather than individuals. In this system, parents produce siblings, who produce cousins, who produce a new set of parents.

Bridewealth is only the beginning. After the girl has matured and the marriage has been consummated, her family eagerly awaits the birth of a child. When this takes place, the payments that a man has been making on his wife's behalf become, instead, payments to ensure the good health of his children.[22] Bridewealth and childgrowth payments thus merge as part of an endless cycle of obligations that a man meets with respect to affinal kin (cf. M. Strathern 1987: 293). Eventually, the newlyweds, now making childgrowth payments, will begin to receive bridewealth on behalf of their own daughters. The cycle of payments thus begins anew, having yielded another generation of social connections.

As I have described them thus far, Kamea marriage payments seem to have little to differentiate them from what is known of affinal relationships elsewhere in the highlands. Yet, what is unique about this system is that it is geared less toward the exchange of brothers and sisters by men and women than it is toward creating gender distinctions in the first place. Marriage and its accompanying system of matrilateral payments begins a process whereby male and female, brother and sister, are engendered against a backdrop of "one-blood" similitude.

Kamea say that girls mature more quickly than boys because they are eager to find a husband and bear children of their own. "Women," I was told, "think only of men and of getting married, this is what makes them grow quickly." The food that the bride's family has been receiving over the years from the groom's family is seen to hasten this maturation process. The idea seems to be that having enjoyed a particularly bountiful diet, the growth rate of a girl rapidly outpaces that of a boy.

It is significant that the main, and certainly the most important, constituent of bridewealth presentations is game—an item that men collect from the surrounding forest environment. Hunting is the productive activity of men sui generis. Although men help their wives with the initial clearing of garden plots, their own productive labor is defined in large measure through their hunting activities. When a couple goes to the bush to stay in their gardens, the daily round consists of women tending to their gardens each morning while men traipse off to the bush to search for game. Indeed, so close is the association between men and hunting that a boy's umbilical cord is cut with <u>pitpit</u>, a type of cane that is used in the manufacture of arrows, while that of a female child is cut with *haka,* bamboo that is used by women for cooking vessels and water containers. Thus the intended productive domains of men and women are inscribed on the body from the moment of birth onward. As one woman explained: "if we cut the navel (*pe'a*) of a boy with bamboo, he would carry a digging stick to harvest sweet potatoes when he grew up. If a girl's umbilicus were cut with <u>pitpit</u> she would always favor hunting over garden work." In the Kamea scheme of things, hunting falls squarely and exclusively within the domain of men.

As noted earlier, bridewealth presentations are given directly to the mother of the girl. This point deserves to be emphasized. Throughout much of highland Papua New Guinea, matrilateral payments are made to a girl's senior male kinsman, either her father or mother's brother, depending on the particular kinship configuration in question. For Kamea, by contrast, these items are given to that very person (the bride's mother) whose own "containing" capacities one hopes to elicit in the daughter. The bride's mother will share the food she receives with members of her immediate family, with one notable exception. Much of the game that is presented to the mother as brideprice is taboo to young boys while they are growing up. (This point is taken up in greater detail in chapter 3.) Therefore, early on a distinction emerges in the eating patterns of "one-blood" siblings: little girls are fed copious amounts of game that is given as bridewealth, while the same food is prohibited to her male "one-blood" siblings. By eating bridewealth, a girl's body comes to contain items of male production (in this case, game), just as she will later contain her husband's sperm and finally any offspring that the marriage produces. A woman's capacity to act as a "container" is gradually developed through these presentations. Pay-

Figure 5. A man distributes pork. Generally, pigs are killed only on special occasions, such as for the payment of bridewealth or during mortuary distributions. Photo by Sandra Bamford.

ments made to a girl's mother become the compelling force that genderizes the female aspect of the "one-blood" tie.

Kamea understandings of childlessness reflect this association between bridewealth and fertility. The men and women of Titamnga contend that it is impossible for a single woman to become pregnant. An unmarried woman can engage in repeated acts of sexual intercourse, but unless bridewealth has been paid in her name, she will never conceive a child. Similarly, barrenness on the part of a married woman is often attributed to the working of a type of sorcery (*pa'a*) by the bride's kinsmen over their displeasure at having received what they perceive to be inadequate bridewealth. Women who fail to conceive a child are often referred to as *oka*—the term for "man," indicating their markedly anomalous place in a world where "femaleness" is defined by the capacity to "contain." From this perspective, a husband and his kin effectively bring about the possibility of the woman's conceiving. By acting as appropriate affines vis-à-vis the parents of the wife, they constitute her capacity to act as a "container," thereby engendering her identity as a reproductively mature female.

We are now in a position to understand what initially struck me as a peculiarity of Kamea kin relations: that sister exchange, although occasionally practiced, does not negate the need to make matrilateral payments.[23] Kamea contend that simply exchanging one's sister for the sister of another man does not cancel out the need to pay bridewealth. Each side to the transaction is still expected to make matrilateral payments, even if the items changing hands are simply identical qualities and quantities of the same food items. In the words of one man: "If my wife's brother married one of my sisters, we would still need to give things to my wife's family. My brother-in-law's side must also give things for my sister. When I saw these things, I would say: 'He gave these things for my sister.' When his sister came to live with me, we would still give things to my wife's mother." To simply exchange sisters is to exchange androgynous beings: persons who are the "one-blood" product of a previous act of procreation, but who have no capacity to move within the world as autonomous female agents.

In her work with the Gawan peoples of Milne Bay Province, Nancy Munn (1986) draws a distinction between siblingship and affinity that is, in many ways, applicable to the situation that I have described for Kamea. Munn notes: "The pre-given identification of cross-sex siblings is that of an undifferentiated whole—an a priori sameness of blood—rather than the unity of two separate, incomplete parts. Thus, opposite-sex siblings must be made separate . . . whereas conversely, marriage makes separate persons into a differentiated whole of complementary parts" (Munn 1986: 41).

If Kamea brothers and sisters represent an undifferentiated whole, homologous by virtue of having been similarly contained, spouses represent, instead, a coming together of two differentiated gendered identities that have achieved separation through intentional human effort. The result of the joining of these differentiated identities is the creation of a child whose own fate depends, in turn, on subsequent acts of differentiation.

Kamea material thus furnishes an important caution concerning the adequacy of orthodox interpretations of kinship and marriage. Lévi-Strauss's (1969) theory of alliance has undergone staunch criticism over the years by feminist writers (G. Rubin 1975), who contend that women in his model become only pieces that are endlessly shifted about by men in a game of social chess. Because women are defined from the outset in terms of their "innate" procreative capabilities, one woman becomes equivalent to any other. Arranging marriage, then, becomes a simple matter of replacing one woman (a sister) with another (a wife). For Kamea, by contrast, the differing values that get attached to wives and sisters means that they *are not* equivalents of one another. A sister may be born with the potential to act as a "container," but a brother cannot activate that potential on her behalf. That task belongs to another man, the man who pays bridewealth for her and who, by this act, precipitates a process of kinship differentiation. Because men and woman are associated with different types of sociality—lineal versus lateral modes of relating, respectively—there are different reproductive consequences for male and female relational configurations.

It follows that for Kamea, siblings constitute a unit that becomes reproductively differentiated through their affinal relations. Giving bridewealth to a woman's kin transforms her by gradually eliciting a female state of being. This act of differentiation continues until there is a complete metamorphosis—the woman is pregnant with a child herself. She has moved from the ungendered state of being "contained"—as represented by the metaphor of "one-bloodedness"—to being able to act as a "container" in her own right. Just as boys "follow their fathers" in terms of the relationships that they form with the land, Kamea girls follow their mothers in terms of their "containing" capacity.

Yet, this identification of a child with the same-sex parent speaks to more than the elicitation of reproductive potential: it also alludes to the way in which offspring relate to same-sex parents and their relatives. Kamea, like many other people the world over, move in a world of multiply constituted kin connections. Individuals can typically trace not one but several possible links to most other persons in the community. A second-degree cross-cousin on the father's side may also be related to a particular individual as a mother's brother. In such cases, a decision has to be made regarding which of these two possible connections to emphasize. Kamea claim that maternal kin would be angry if all children in a given sibling set used the father as a point of reference in sorting out ambiguous cases. The perceived difficulty is solved by having girls—particularly as they get older—rely on their mother as the determining link. As one of my consultants explained: "Girls belong to their mother's line, boys to their father's."

The Kamea world, then, becomes split: girls identified with their mothers and boys with their fathers. However, it is important to note that the association between a mother and daughter goes back no further than one generation and does not form the basis of long-standing temporal connections. A daughter may follow her mother in classifying kin, but that mother will follow her own mother, who follows her own, and so on. There is no sense in which female sociality is transmitted through time in such a way as to make time evident as a lineal flow across the generations. Instead, for women, time consists of a series of substitutions. In this way, female sociality differs from male forms of relatedness (see chapter 1). Men draw upon land and nonhuman resources to invest their sociality in an externalized world. This gives male relatedness a temporal dimension that women's relationships lack. Female sociality is confined to a single generation, both in terms of lateral relationships—"one-blood bonds"—and in the nature of those ties that bind a mother to her daughter.

CYCLES OF TRANSFORMATION

I have argued that Kamea marriage arrangements are focused as much on the creation of gender as they are on inaugurating a procreative relationships. Cross-cousins are central to this process of differentiation. By giving and receiving payments on their son's and daughter's behalf, they begin a process whereby siblings—beings of undetermined gender—come to be defined as male and female agents, with a particular kind of reproductive potential. Yet cousins do more than elicit gendered states; they also engender different modes of relating. As Yanagisako and Collier (1987) have argued, gender and kinship are not separate analytical domains, but mutually constitutive components of a social whole. Through their "work" in creating difference from a prior state of unity, cousins also assign different meanings to the husband-wife and brother-sister relationship, despite what Kamea see as being their underlying similarity.

That siblingship and affinity are analogues of one another is part of a conscious model that Kamea themselves hold of the world. I first became aware of the perceived equivalence between these two types of relationships about a year after taking up residence at Titamnga. One afternoon, I was returning to the village after having spent several hours in the company of Eniyi, one of my oldest consultants and a renowned expert on Kamea myth. As we walked along the footpath that connected the two villages, I idly expressed to Kawa, my traveling companion, my sense of frustration at not having fully mastered the Kamea language. It had seemed to me that at one moment Eniyi was saying that the Sun and the Moon—two important figures in Kamea mythology—were related as husband and wife and the next that they were related as brother and sister. Kawa stopped in his tracks and beamed at me explaining that that was, in fact, what Eniyi had said. When I asked for clarification, Kawa responded by saying that the Sun and the Moon were husband and wife, brother and sister at the same time (cf. Godelier 1986: 36; Lévi-Strauss 1976: 216).

The implicit sameness that underlies siblingship and affinity in myth has terminological correlates in the lives of actual people. As I struggled to make sense of what Eniyi and Kawa had told me, I began to pay closer attention to how husbands and wives addressed one another. Although a husband may refer to his wife as *nka apaka* ("my woman") and a woman to her husband as *nka oka* ("my man"), an equally common practice is for the spouses to refer to one another using terms for siblings. Thus, a wife might call her husband *nka dato* ("my brother") and a husband might address his wife as *nka nabi* ("my sister"). This same practice applies to third-party interactions. A man meeting a married couple in their garden may greet them, "Afternoon, brother and sister." As Oates and Oates (1968) note of the Kamea language more generally: "When speaking of one's own or one's hearer's relationships, no distinction is made between the spouse and the sibling relationship" (169).

The substitutability of sibling and spousal terms highlights the contrived nature of sociality for Kamea. Neither kinship nor gender are essentialized identities; rather, they are elicited on an ongoing basis through conscious human effort (cf. Wagner 1975, 1977). Siblingship and affinity are identical relationships in this world, differentiated only by their polarized gender attributes. Cousins are significant because they create those distinctions upon which Kamea social life depends. Just as male and female pass through differentiated and undifferentiated states (see Bamford 1997, 1998b, 2004), so too do siblingship and affinity contain, at least implicitly, the possibility of merging as undifferentiated relationships. Cousins act as the pivot between these two sets of distinctions and keep certain types of people and relationships separate until they can productively be brought together again.[24]

By virtue of their functioning as a nexus of transformation in Kamea social and cultural processes, cross-cousins hold a particularly compelling place in Kamea cultural thought. In chapter 4 I will discuss the nature of this relationship in greater detail. Here, we need only note that the relationship between cross-cousins embodies the seemingly contradictory states of intimacy and avoidance. On the one hand, cross-cousins are the only category of kin that Kamea explicitly relate to an image of the human body. My friends at Titamnga told me, "My cousin is my nose," a point to which I shall return when discussing Kamea constructions of death. Additionally, cross-cousins are said to enjoy a particularly close relationship with one another. Should an altercation develop, for example, between one's father and one's cross-cousin, it is to the latter that one is expected to provide support and loyalty. Finally, I should note that cross-cousins are expected to act as undertakers for one another, an obligation that in the past meant sitting underneath the corpse of one's dead kinsperson as it decomposed and smearing oneself with the fluids of decomposition that dripped from the body.

On the one hand, then, the identification between cousins is so intense that they appear to be extensions of the self. At the same time, however, the relationship between them is hedged with a dizzying array of taboos and avoidances. Unlike any other category of kin, interactions between cousins are subject to numerous interdictions. One should avoid stepping over the legs of one's cross-cousin or walking behind them when they are seated on the ground. Similarly, one must avoid uttering the proper name of one's cross-cousin out loud and should refer to him or her instead by euphemisms or through the categories set out in the kinship terminology (i.e., *ndawo* for a male cross-cousin, *nawi* for a female one). Failure to adhere to these restrictions will cause the children of one's cross-cousin not to develop properly, as the following excerpt illustrates:

S: If your cousin was sitting down, could you step over them?

M: No. It is my cousin. I must excuse myself first. It would not be good for me to step over them. If I stepped over my female cousin and she had a baby, she would not give birth to a nice fat child. She would carry a <u>bun nating</u> (a child with no meat on its bones). As for my female cousin, she would be unable to produce breast milk.

S: Could you step over a male cousin?

M: No. The same thing would happen. When he and his wife had children, they would be <u>bun nating</u>. The children would remain small and stunted and fail to grow.

Although cross-cousins are seen to be closely linked in Kamea thought, these restrictions ensure that a degree of separateness between them is maintained. Cousins both unite and divide at the same time, an appropriate analogue to the role they play in the kinship system (see chapter 4 for further details).

NONEMBODIED CONNECTIONS

This chapter opened with a consideration of Dame Butler-Schloss's ruling concerning the case of the "black twins" in the United Kingdom. As difficult as this case was to resolve on legal grounds, commentators (Birkett 2002; Dyer 2002; Verkaik 2002) agree that the situation would have been far worse had the "mix-up" involved the eggs of Mrs. A, and not just the sperm of Mr. B. Had this transpired, the courts would have been forced to negotiate the rights of not just two parties, but three: the sperm donor, the egg donor, and the woman who gave birth. Under these circumstances, Mrs. A could have staked a claim to the children as their gestational mother and an already laborious decision-making process would have been made even more so.

To the extent that the new reproductive technologies are "new," what is novel about them is that they have teased apart elements of what was once understood to be an integrated whole—sexual intercourse, gestation, and childbirth. Today, these components are routinely disassembled in laboratory settings and are mixed with the "substance" of an expanding market economy.

While the case of the "black" twins privileged genes as the foundation of intergenerational continuity, "biological" connectedness is sometimes construed to rest on different grounds. In an illuminating paper that deals with these themes, Charis Cussins (1998) recounts her experiences in carrying out ethnographic research at infertility clinics across North America. Several couples that she interviewed were undergoing donor egg in vitro fertilization; a procedure in which the maternal genetic material is contributed by an egg that is derived from the ovaries of a woman who is not intending to gestate or raise the child herself. In describing the experiences of one of her consultants who was seeking to become pregnant through these means, Cussins writes: "Giovanna accorded her gestational role a rich biological significance: she said that the baby would grow inside her, nourished by her blood and made out of the very stuff of her body, all the way from a two celled embryo to a fully formed baby. . . . Giovanna cast her gestation in biological terms, appealing to blood and shared bodily substances" (Cussins 1998: 44). A similar reading of biology informed the now famous court case, *Johnson v. Calvert,* in which a gestational surrogate (Anna Johnson) petitioned for parental rights and custody to an unborn child that had been conceived through the gametes of Mark and Crispina Calvert, claiming that she had formed an attachment to the child while gestating it (Hartouni 1997: 85).

However different these two cases may seem from that involving the "black twins," they share in common one important similarity. Parentage is seen to rest on a bond of corporeal attachment, whether established at conception or by virtue of pregnancy and birth. Put otherwise, what is taken as axiomatic in all of these discussions is that an embodied link of some kind grounds intergenerational continuity.

The world of assisted reproductive technologies provides an illuminating vantage point from which to reflect upon anthropological theories of kinship. Nearly forty years ago, David Schneider (1968; 1984) cautioned anthropological field-workers to put their own assumptions concerning relatedness into interpretative abeyance. Central to Schneider's critique of existing kinship theory was a questioning of the extent to which sexual reproduction could be taken as a universal feature upon which human social systems were built. In Euro-American formulations, consanguinity has to do with the reproduction of human beings, and reproduction in turn is understood to be a sexual and biological process. In the folk wisdom of the West, sexual reproduction creates physiological links between human beings and these are understood to have important attributes apart from any meanings that might get attached to them (Schneider 1984: 188). Schneider challenged the universality of these assumptions. He asked us to take kinship as an "empirical question"—a hypothesis to be tested—rather than to assume from the outset that it had cross-cultural significance (Schneider 1984: 200). In the wake of his work, anthropologists could no longer take the "facts of life" for granted, a finding that similarly called into question the invariant nature of the genealogical grid.

The past decade has witnessed the emergence of a "new" theoretical model that attempts to take Schneider's insights to heart. The currently dominant paradigm (Carsten 1995, 1997, 2000; Rival 1998; Stafford 2000; Weismantel 1995a, 1995b) strives to move beyond understanding relatedness in terms of a distinction between "social" versus "biological" relationships. In this model, "physical" ties are not given at birth, but are created through intentional human effort (Viveiros de Castro forthcoming). As we have seen, proponents of this model strive to highlight how commensality and co-residence can combine to engender consubstantial relations. This paradigm thus emphasizes the various means by which corporeal relations can be created between persons after birth.

While these newer studies have helped to challenge the association between parentage and physiological reproduction, they continue to take as axiomatic the idea that cross-generational relations necessarily entail an embodied and substantial link between parents and children. Implicit in this approach is the idea that Western notions of hereditary substance can have only one antithesis, namely, substance acquired not at birth but processually over time and as a consequence of human effort. What also remains unchallenged is the view that the parent-child tie serves as the generative nexus of social life and that procreative kinship has to do with the production of individual bodies and persons rather than sets of relations.

Within this context, Kamea represent a particularly compelling case, providing us with a glimpse into a world in which the parent-child tie is not conceived in genealogical terms. Although the men and women of Titamnga are highly articulate when it comes to espousing their views on what goes into the making of a baby, conception is not used as a means of tracking social relations through time. While both a mother and a father are understood to contribute bodily substance to the making of a child, this is seen to be a rather unremarkable feature of the ensuing relationship. My own efforts to ground motherhood and fatherhood in a substance-based link met with everything from amusement to total indifference and incomprehension. Physiology, even as it is indigenously understood, is of little use in grasping the dynamics of Kamea social life. In order to understand what connects and disconnects people in this world, it is necessary to move from a substantive to a relational point of view.

In this and the previous chapter, I have described Kamea social life in terms of a contrast between male and female relational configurations. Men and women each engender their own modes of relating that can be glossed in terms of a contrast between lineal and lateral modes of relating. Relationships defined through women are of a short temporal duration (one generation only), and receive their most potent expression in the horizontal ties of "one-blood" siblings. Kamea men, by contrast, are implicated in the definition of relationships through time. A boy creates his own place in the nexus of remembered social relationships by virtue of the ties that he forms with the land and a variety of nonhuman resources. Men have nothing to do with the definition of horizontal relations, which are based on women and their capacity to contain. Kamea women, for their part, while providing the foundation upon which lateral relationships are based will never serve as points of attachment in tambuna storis (see chapter 1). It is the intersection of these two distinct relational forms (rather than a reproductive bond as it is understood by Euro-Americans) that gives Kamea social life its forward-going momentum.

Significantly, neither male nor female relational modes are conceptualized in terms of the transmission of substance, and neither is seen to result in an embodied connection between parents and offspring. This carries with it a number of important implications in terms of how persons, bodies, and time are perceived. The generation of social life does not unfold along a unidirectional temporal sequence, as Westerners understand it; instead, it entails the cycling back and forth between states of unity and disjuncture. Cousins are central to this process of differentiation. By initiating affinal relationships between their children, they begin a process through which gendered forms are created and productive cross-sex relations are differentiated from unproductive ones.

What Kamea have to tell us about human social life is of no minor importance. Over the last few years, the study of kinship has taken on a new urgency, propelled in part by changing definitions of the family in Europe and North American. With the growth of new reproductive technologies and the rise of gay and lesbian studies, members of our own society are being called upon with ever increasing frequency to rethink our own ideas about what is innate and given with respect to human relatedness. In a world where embryos can be "put on ice" and the dead can be forced to procreate, Kamea furnish us with a new perspective upon which to reflect upon what is essential about persons, reproduction, and cross-generational relationships. They provide us with a glimpse into a world in which kinship is not understood to entail an embodied connection—either created intentionally (as the newer processual model holds) or given at birth (as traditional studies of kinship would have it). More specifically, they allow us to see that significant social ties may be based on more than bodily substance, eventuating also from nonembodied connections.

THREE

Embodiments of Detachment

In October 1986, the San Diego District Attorney's Office issued a court order for Pamela Rae Stewart to appear before the bench. Stewart, a twenty-seven-year-old mother of three, was charged with criminal neglect for contributing to the death of her newborn son by failing to follow her doctor's advice before the infant's birth.

The case against Pamela Rae Stewart is believed to have been the first in North America in which a woman was charged with criminal liability for her conduct during pregnancy.[1] According to court testimony, Ms. Stewart suffered from a medical condition known as *placenta previa,* in which the baby cannot descend through the birth canal without detaching the placenta from the uterine wall, threatening a maternal hemorrhage and depriving the baby of oxygen, sometimes fatally (Begley et al. 1986). Stewart's doctor had advised her to refrain from using street drugs and having sexual intercourse, and instructed her to seek immediate medical attention if she began to hemorrhage. Police allege, however, that on the day that Stewart delivered her child, she took amphetamines, had sex with her husband, and did not call paramedics until twelve hours after she began to bleed. Her son, Thomas Travis Monson, "was born brain dead with amphetamines in his system" (Begley et al. 1986). He died six weeks later while in foster care.

During the trial, San Diego Deputy District Attorney Harry Elias claimed that Stewart was indicted because she had failed to follow doctor's orders during her pregnancy, which resulted in "injury to her and injury and death to her baby" (quoted in Cassens 1987a). More specifically, he charged Stewart with failing to "furnish necessary medical attendance" to the fetus under the California Penal Code, Section 270, a statute typically used to collect child support (Cassens 1987a). Section 270 calls for a sentence of up to one year in jail and up to a two thousand dollar fine for any person who "willfully omits without lawful excuse to furnish necessary clothing, food, shelter, medical attendance, or other medical care for his or her child" (Anonymous 1988: 994). The statute goes on to say that "a child conceived, but not yet born is to be deemed an existing person insofar as this section is concerned" (Anonymous 1988: 994).

While the California courts eventually dismissed the case against Stewart, more recent defendants in "fetal abuse" cases have not fared as well.[2] In 1989, a Florida judge sentenced Jennifer Johnson to fifteen years probation on the conviction of "delivering illegal drugs via the umbilical cord" to her two babies (Bordo 1993: 82). The cocaine-addicted mother (whose children were in good health) could have received a thirty-year jail sentence under provisions of a Florida law usually applied to drug dealers (McNamara 1989). Later that same year, a Massachusetts woman who miscarried in the wake of an automobile accident in which she had been driving intoxicated was prosecuted for "vehicular homicide" of her fetus—a charge that carries with it a mandatory jail term of one year (McNamara 1989). In Canada, a woman was charged with "willfully inflicting F.A.S. (Fetal Alcohol Syndrome) upon her child," by refusing to seek help for her alcoholism after a doctor suggested she do so (Balisby 1987: 1225). A Connecticut woman was charged with endangering her fetus by swallowing cocaine as police attempted to arrest her for drug possession. In Wisconsin, a sixteen-year-old pregnant girl was held in secure detention for the sake of her fetus because she "tended to be on the run" and to "lack motivation or ability to seek prenatal care" (Kolder, Gallagher, and Parsons 1987: 1195). Diane Pfannenstiel was jailed in Wyoming and charged with the "crime" of "drinking while pregnant" when she presented herself at a local hospital for treatment after having been the victim of spousal abuse. Pfannenstiel lost custody of her two children and was prosecuted by the courts for felony child abuse (Goodman 1990). In the midwestern United States, a court of appeals upheld an automotive battery plant's seven-year-old "fetal protection policy" preventing "fertile" women (in effect, all women) from taking jobs that would expose them to lead. In this particular case, the court ignored testimony concerning the individual reproductive plans of female employees—many in their late forties, or with completed families—and testimony indicating that lead poses a similar danger to the reproductive health of men. Instead, it accepted testimony that said making the workplace safe would be too expensive to be a viable option (Pollitt 1990). In South Carolina, a dozen women were arrested after the hospital they went to for maternity care tested them for drug use and turned them in to the police on "fetal abuse" charges (Goodman 1990).

As the foregoing suggests, the last few decades have witnessed the growth and proliferation of a doctrine of "fetal rights." Precipitated, in part, by the development of new technologies that make prenatal evaluation and treatment of the fetus more accessible (Rhoden 1986: 1951), courts throughout Europe and North America are in the process of assigning an elevated moral and legal status to the fetus, granting it independent "personhood" apart from the woman who carries it. Claims granting legal status to the unborn are based on scattered case and statutory laws that assign implicit legal recognition to the fetus (Gallagher 1987: 11). Various legal precedents have recently been set in this regard. If doctors suspect an unborn child to be at serious risk, for example, a pregnant woman may be forced to submit to a cesarean section despite her explicit refusal. It has also been demonstrated that in the U.S. legal system, "government restraints may be placed on a pregnant woman's physical activities, diet and lifestyle, mothers can be held liable in tort for injuries to children occasioned by their 'prenatal negligence,' and that terminally ill pregnant women [can] be excluded from the protection of 'living will' statutes" (Gallagher 1987: 11). John Robertson, a leading proponent of fetal rights legislation, argues for a theory of "contingent legal personhood" that would subject women to "retrospective criminal and civil liability" for all "damaging acts and omissions" they commit before a child is born. Both he and other advocates of fetal rights have been calling for the development of far-reaching state interventions that would monitor the lives of pregnant women. More specifically, Robertson imagines a world in which pregnant women

> may be prohibited from using alcohol or other substances harmful to the fetus during pregnancy, or be kept from the workplace because of toxic effects on the fetus. They could be ordered to take drugs, such as insulin for diabetes, medications for fetal deficiencies, or intrauterine blood transfusions for Rh factor. Pregnant anorexic teenagers could be force-fed. Prena-

tal screening and diagnostic procedures, from amniocentesis to sonography or even fetoscopy could be made mandatory. And, in utero surgery for the fetus to shunt cerebroventricular fluids from the brain to relieve hydrocephalus, or to relieve urethral obstruction of bilateral hydronephrosis could also be ordered. Indeed, even extra-uterine fetal surgery, if it becomes an established procedure could be ordered, if the risk to the mother were small and it were a last resort to save the life or prevent severe disability in a viable fetus. (Robertson, quoted in Rhoden 1986: 2027)

Sam Balisby (1987) and Margery Shaw (1984) similarly champion the development of far-reaching fetal rights legislation. They would go even further than Robertson, by calling for compulsory screening of alcoholic or drug-addicted mothers and taking physical custody of a pregnant woman to protect her fetus from the "unhealthy" lifestyle choices. In the case of repeat offenders, they advocate the termination of parental rights.

As might be expected, the development of fetal rights rhetoric has prompted a strong reaction on the part of feminist scholars in Europe, Australia, and North America (see, for example, Bordo 1993; Davies and Naffine 2001; Gallagher 1987; James and Palmer 2002; Kolder, Gallagher, and Parsons 1987; Rhoden 1986). Critics have argued that the emergence of laws intended to protect the fetus threaten to pit the interests of women against those of unborn children, thereby creating an adversarial relationship between a woman and her child.

In the following section, I set out the contours of this debate. My intention is not to take a particular stand in this controversy; nor will I be able to summarize the full range of positions that different scholars take. Instead, I will use the emergence of fetal rights rhetoric, and the reaction against it, as a means of highlighting certain ideas that underlie how Europeans and North Americans think about bodies, persons, gender, and agency. I shall argue that although the positions of advocates and opponents are at one level radically opposed (and thus hold out radically different implications for women), from another perspective, they share in common several important ideas that derive from their mutual adherence to a biological paradigm. Having highlighted those assumptions that frame this controversy, this conceptual system will then viewed through the defamiliarizing lens of Kamea ethnography. This chapter thus serves to highlight the extent to which a genealogical framework carries with it a particular vision of personhood and embodied experience.

In her paper "Are Mothers Persons?" Susan Bordo highlights what is seen by many feminist scholars as a fundamental contradiction of fetal rights legislation: "Although law and medicine claim to have a unified and coherent tradition concerning individual rights, in fact, two different traditions have been established, one for embodied subjects, and the other for those who come to be treated as mere bodies despite an official rhetoric that vehemently forswears such treatment of human beings" (Bordo 1993: 72). Put somewhat differently, the legal tradition of Europe and North America displays a troubling double standard. On the one hand, common law has long recognized that each individual has a right to control his or her own person and a concomitant right to be free from nonconsensual invasions of bodily integrity (Nelson, Buggy, and Weil 1986: 746). The concept of the free, autonomous, and equal person is fundamental to modern liberal legal systems (Davies and Naffine 2001: 5). At the same time, however, this right is regularly repealed in the case of a pregnant woman whose physical boundaries are often policed against her (Savell 2002: 40).

Critics of the fetal rights position base their case on the fact that our legal tradition places an extremely high value on self-determination and bodily integrity. As the United States Supreme Court acknowledged over a hundred years ago: "No right is held more sacred, or is more carefully guarded by the common law, than the right of every individual to the possession and control of his own person, free from all restraint or interference of others, unless by a clear and unquestionable authority of law. As well said by Judge Cooley, 'The right to one's person may be said to be a right of complete immunity; to be let alone' " (*Union Pacific Railway v. Botsford* [1891], quoted in Bordo 1993: 72).

Indeed, the right to bodily integrity is valued to such a degree that judges have consistently refused to force individuals to submit without their consent to medical procedures, even though the life of another may be endangered as a consequence (Bordo 1993: 73). The 1978 case *McFall v. Shimp* dealt with the issue of compelling a competent adult to "donate" bodily tissue against his will. Robert McFall, a victim of aplastic anemia, sought a court order that would have required his first cousin, David Shimp, to undergo testing for compatibility of bone marrow and to donate tissue if sufficient compatibility were discovered. "The transplant procedure would have offered the plaintiff at least some hope of survival while the risk of harm to the unwilling donor was small to nonexistent. The court dismissed the request of the plaintiff, finding no moral duty on the part of the defendant to provide assistance" (Nelson, Buggy, and Weil 1986: 755). In rendering his decision, Judge John Flaherty defended his ruling in the following terms: "The common law has consistently held to a rule which provides that one human being is under no legal compulsion to give aid or take action to save that human being or to rescue. For our law to compel the defendant to submit to an intrusion of his body would change the sanctity of the individual and would impose a rule which would know no limits and one could not imagine where the line might be drawn" (Gallagher 1987: 26).

Two examples of similar suits that have been equally unsuccessful include a Seattle woman who sued her four-year-old son's father to donate marrow for her son, who was suffering from leukemia; and an Illinois man who sued the mother of his son's twin half-siblings, to have tests done to see if they could be marrow donors for his son (Bordo 1993: 73). Neither petition for assistance was granted. As a recent survey of the law of Samaritanism concluded: "the basic and well-established common law principle is that one individual is not required to volunteer aid to another" (Gallagher 1987: 24).

The doctrine of informed consent (an equally important tenet in the Euro-American legal tradition) is intended to protect the rights of all citizens to self-determination. At the root of this principle is the idea that the body cannot be treated as inert matter; rather, it must be viewed as "invested with personal meaning, history and value that are ultimately determinable only by the subject who lives 'within it'" (Bordo 1993: 74). In keeping with the value of "informed consent," even when it is in the "best interest" of the patient, competent individuals can refuse medical treatment, even if this will result in their death.

The right to refuse treatment is grounded in common law "and is encompassed within the constitutional right to privacy" (Rhoden 1986: 1971). This right is fundamental and, at least in principle, is rarely contravened. "Refusals have been upheld even when the patient was mentally ill" (Rhoden 1986: 1971). For example, judges of Massachusetts's highest court have banned "the nonemergency use of antipsychotic drugs on institutionalized mental patients who refuse them, unless the state proves both that the patient is incompetent to make such a decision and that, if competent, the patient would choose to take the drug" (Gallagher 1987: 20). As Rhoden notes, the doctrine of informed consent illustrates two important principles of Anglo-American law: "First, it shows the extent to which patient autonomy is respected and highlights how nonconsensual surgery deviates sharply from our society's respect for such autonomy. Second, it shows that risk benefit judgements about medical treatment are the patients' to make and not the doctors'" (Rhoden 1986: 1970). Deliberate disregard of a patient's refusal is tantamount to assault and/or battery in the eyes of the law (Kolder, Gallagher, and Parsons 1987: 1195).

As opponents of fetal rights legislation are quick to point out, so central is the legal tradition of respect for autonomy and self-determination that these principles are revoked only under the most exceptional circumstances. In *Rochin v. California* (1952), police broke into the room of a person suspected of selling narcotics and forced him to regurgitate two capsules he had swallowed (Bordo 1993: 74). Although the officers involved in the case were not charged with a crime, the court registered its "strong disapproval of these actions" (Rhoden 1986: 1983; cf. Bordo 1993: 74–75). Even with suspected criminals, however, the law draws the line at surgical intervention. In *Winston v. Virginia* (1985), law enforcement authorities hoped to obtain a bullet that was lodged in a defendant's chest as evidence of criminal wrongdoing. The proposed surgery required general anesthesia and there was some uncertainty concerning the size of the incision that would need to be made (Rhoden 1986: 1985). The Supreme Court ruled that surgery would violate the defendant's constitutional rights and said that the "overriding function of the fourth amendment is to protect personal privacy and dignity against unwarranted intrusion by the state," and that "these values were 'basic' to a free society" (Gallagher 1987: 21). The distaste that Europeans and North Americans evince for violating the right of self-determination extend even to dead bodies. Cadavers are legally protected against use of their organs and tissues without explicit prior consent, even when their use could save the lives of others (Bordo 1993: 311).[3]

As several feminist scholars have argued, while self-determination and nonsubordination emerge as cardinal values in Europe and North America, this right is often suspended in the case of pregnant women. While it may be battery to administer medical treatment to nonconsenting competent adults (even if the subject in question will die without it), pregnant women appear to be subject to an entirely different set of legal edicts. Beginning in the early 1980s and continuing until the present day, there has been a dramatic rise in the incidence of court-ordered obstetrical interventions. In 1987, 47 percent of the obstetricians surveyed by the *New England Journal of Medicine* agreed that the precedent set by courts in cases requiring emergency cesarean sections for the sake of the fetus should be extended to include other procedures, such as intrauterine blood transfusions (Bordo 1993: 80), if necessary to preserve fetal well-being (cf. Kolder, Gallagher, and

Parsons 1987: 1192). In one of the most controversial cases involving a forced cesarean section, George Washington University Hospital asked a court whether it could intervene, performing a cesarean section on Angela Carder, seriously ill with cancer, against her expressed wishes and those of her husband, her parents, and her doctors (Bordo 1993: 77). Acknowledging that the operation would probably shorten her life without saving the life of her fetus, the judge nonetheless signed the order. The operation was performed before the mother's lawyers could appeal. Ms. Carder died, as did her unviable fetus (Pollitt 1990: 409).

The Carder case was not unprecedented. As early as 1964, the New Jersey Supreme Court recognized the rights of an unborn child in the case of *Raleigh-Fitkin-Paul Morgan Memorial Hospital v. Anderson*. "The suit was initiated by a hospital seeking an order to compel a Jehovah's Witness to submit to blood transfusions in her eighth month of pregnancy. The woman, in danger of hemorrhaging, refused the transfusions on religious grounds. In granting the hospital's request, the court stated that 'the unborn child is entitled to the law's protection and that an appropriate order should be made to insure blood transfusions to the mother in the event that they are necessary in the opinion of the physician in charge at the time'" (Balisby 1987: 1229). In a similar case, in the District of Columbia in 1986, when Ayesha Madyun refused a cesarean on religious grounds, the judge ruled that *not* to issue a court order forcing her to have the operation would be tantamount to "indulg[ing]" Madyun's "desires" at the expense of the fetus (Bordo 1993: 78). The judge went on to say: "It is one thing for an adult to gamble with nature regarding his or her life; it is quite another when the gambler involves the life or death of an unborn infant" (quoted in T. Lewin 1987: 1).

On the basis of what is argued to be a fundamental discrepancy between law and legal practice, a growing assault against fetal rights legislation is emerging. Many feminist scholars and legal theorists are quick to point out that pregnant women are regularly denied rights that are normatively extended to criminals, mental patients, even the dead:

> If our society will not compel someone to undergo a bodily invasion or tissue transplantation for the benefit of another, how can society view pregnant women refusing treatment any differently? The basic values at stake are the same: the freedom to choose one's destiny and to maintain one's bodily integrity. . . . [C]ompelled treatment cannot plausibly be justified by the need to save the fetus' life: We do not legally compel others to save

the lives of live-born persons, even when the action required is much less burdensome or invasive. (Nelson, Buggy, and Weil 1986: 755)

Several commentators in this debate have also argued that fetal rights legislation carries with it a "slippery slope" of unintended consequences. One can imagine, for example, that prenatal liability need not be restricted to pregnancy; it could be extended back into the indefinite past, before conception. As Gallagher points out, passing on "unhealthy genes" could come to be viewed as tortious behavior. "Regulation could become so particularized that women at high risk of having a child with neural tube defect could be required to decrease that risk by taking folic acid supplements during the last two weeks of every menstrual cycle" prior to actual conception (Gallagher 1987: 44). What is perceived to be at stake is the denial of a pregnant woman's subjectivity. According to Bordo, the fetal rights position is one in which "pregnant women are not subjects at all (neither under the law nor in the zeitgeist) while fetuses are *super*-subjects. It is as though the subjectivity of the pregnant body were siphoned from it and emptied into fetal life" (Bordo 1993: 88).

The critique of fetal rights legislation highlights an important tenet of Euro-American thought: namely, the equation of personhood with self-determination and bodily integrity. What opponents are asserting is not simply that the emerging law treats women unfairly; rather, critics are arguing that these laws strip women of their very subjectivity by depriving them of the basic right to manage their own bodies. From this perspective bodies are, or at least should be, under the exclusive proprietal ownership of a singular and autonomous self. Indeed, it is this right to self-proprietorship that is seen to define legal subjectivity in the West.

The position taken by supporters of these laws reveals an equally telling range of ideas. Proponents do not debate the principle of bodily integrity (this is taken as axiomatic); instead, they claim that pregnant women are somehow "different":

A pregnant woman is in a unique situation. If a woman refusing medical treatment were not pregnant, only her interests would be at stake. During pregnancy, however, the woman literally encloses another being, the fetus, to which she is directly, physically and intimately attached. What she does, or fails to do, can have an immediate physical effect on this other being in a manner truly singular in human experience. As long as the fetus remains within her, the pregnant woman is connected to that being in a way that no other person is connected to any living person. *In contrast, after the fetus is born and the two are separated,* the mother could refuse life-saving treatment and leave the baby motherless, but she could not take her child to the grave with her. (Nelson, Buggy, and Weil 1986: 720; italics added)

In staking their claims, fetal rights advocates couch their argument in a language that promotes the moral status of the fetus. If it can be shown, they reason, that the fetus is a "person," then it follows that it is worthy of being granted rights, including the right to be "born of sound mind and body" (Nelson, Buggy, and Weil 1986: 735). In making their case, proponents—and increasingly courts—draw upon the landmark abortion decision in *Roe v. Wade* (1973), where biological viability is used as the defining criteria of personhood. In particular, fetal rights advocates rely on the following passage from *Roe,* in which the originating moment of personhood is not birth, but rather the point at which an unborn child is capable of independent existence. "With respect to the State's important and legitimate interest in potential life, the 'compelling' point is at viability. This is so because the fetus, then, presumably has the capability of meaningful life outside the mother's womb. . . . If the state is interested in protecting fetal life after viability, it may go so far as to proscribe abortion during that period except when it is necessary to preserve the life or health of the mother" (Gallagher 1987: 15). From the perspective of fetal rights advocates, the capacity to be born alive—separate and independent from one's mother—is the condition of legal personhood. Separation, distinction, and independence figure centrally in the definition of Western selfhood (Savell 2002: 37). From that point in time when a child can exist outside the womb as an independent entity, the mother-to-be is legally charged with the task of protecting the fetus, and treating it as an entity with full rights, equal to those of the born. A fetus of twenty-eight-weeks gestation is presumed to be a child capable of being born alive, with the result that it is protected under the Life (Preservation) Act, which limits the conditions under which a woman could seek an abortion. The capacity for independent biological existence, then, is used as the basis upon which legal rights are assigned.

Some fetal rights advocates would go one step further and use the viability line established in *Roe vs. Wade* as an element in a waiver argument, asserting that once the woman "decides to forgo [*sic*] abortion and the state chooses to protect the fetus, the woman loses the liberty to act in ways that would adversely affect the fetus" (Robertson, quoted in Gallagher 1987: 31). From this point on, the mother's duties to protect the fetus from harm actually *increase* because she chose not to exercise her right to have an abortion during the first trimester. Under these circumstances, it is argued, the woman should be held accountable for all acts, those of omission and those commission (i.e., taking drugs, failing to seek a prenatal advice from a trained practitioner of biomedicine, etc.), that are carried out during the course of her pregnancy (Begley et al. 1986: 756; cf. Gallagher 1987: 11).

A MELANESIAN PERSPECTIVE ON BODILY INTEGRITY

Both sides of the dispute concerning fetal rights bring to the fore an important feature of Western thought, namely, the degree to which images of personhood are fueled by a concept of the possessive individual, particularly the notion that persons are free and autonomous agents, fully in charge of their body and all its experiences. As Janet Gallagher (1987) has written: "Fetal rights advocates assert that the courts' expansion of causes for action for wrongful death for injuries sustained in utero indicates a trend toward full legal status for the fetus. *They maintain that the determinative issue in such cases has been the fetus' biologically independent existence.* In this view, the courts have replaced the traditional birth requirement for tortious liability with a new viability line that recognizes the fetus' genetic individuality" (38; italics added).

While advocates and opponents take markedly different stands on the issue of fetal rights, several common assumptions underlie their respective positions. First is the notion of "bodily integrity"; the idea that bodies are (or should be) autonomous, self-enclosed, and self-directed forms. Second, is the axiom that except under very "unusual" circumstances (such as pregnancy), an individual exercises sole proprietal rights over his or her body. Third, is the idea that that a mother and unborn fetus operate as a singular embodied entity up until the moment of birth—after that, they are fully separate beings.

In the pages that follow, I explore an alternative cultural logic. Through an examination of how food taboos intersect with the activities of the men's cult, I show how Kamea male initiation can be seen as a process of "decontainment," wherein a composite social form, a mother-son dyad, is differentiated to yield two separate and autonomous beings. Far from being given in the "nature of things," Kamea bodies do not exist as intrinsically differentiated entities, but pass back and forth between singular and composite states. One key way in which bodies are brought together in social relations is through being gendered—a process that eventuates at least in part from the performance of the initiation-taboo complex. By examining Kamea notions of connectedness and disconnectedness, I document how agency (the capacity to act as an autonomous and gendered being) is created in a world where identities are perceived in nonbiological terms.

In pursuing this theme, I return to classic themes in the ethnographic literature. Myer Fortes once wrote of the privileged place that food taboos occupy within the discipline of anthropology. Eating, he argued, is a uniquely individual act in that each person must eat for herself or himself—it is not an activity that one person can undertake for another. Furthermore, eating is both organic and social: "it is a means by which we are not merely made aware of an external reality but take permitted parts of it into our [body]" (Fortes 1966: 16). This being so, it is not surprising that food taboos have captured the attention of anthropologists. Not unlike Lévi-Strauss's (1969) leitmotif—the incest prohibition—dietary restrictions intrigue us with the possibility of casting light on some of the West's most persistent analytical dilemmas—the relationship between mind and body, and individual and society.

I am not going to offer up a new theory of taboo in this chapter. Nor will I enter into some of the more specific debates that have surrounded how anthropologists have approached the subject (see Lévi-Strauss 1966). Instead, I want to examine one feature of taboo that has received scant analytical attention: how taboos reflect values of "sameness" and "difference" at the same time. Most anthropological treatments of taboo have taken the issue of boundary maintenance as their point of departure. Taboos, we are told, are about preserving the integrity of a classifactory system; they have to do with keeping certain categories of people and things apart (Douglas 1966; E. Leach 1964: 210; Schieffelin 1976: 65–70). Pollution results when the dividing line between classes have been breached; when what is distinct is threatened by contagion to become an undifferentiated mass. To follow a taboo, then, is to shore up a system of cultural differences and to prevent connections from forming between cultural domains that must not be mixed.

Among Kamea, taboos frequently *do* mark off categories of persons and things. But they are also a means by which such distinctions are created in the first place, and they furnish a means by which such acts of differentiation come about. Scholars of Melanesia have long noted that the work of most New Guinea cultures is not centered on reproducing a formal model of society, but rather, is geared toward creating particular kinds of social relationships (M. Strathern 1988; Wagner 1975, 1977). This work is ongoing in the sense that what was once differentiated inevitably slips through time back to a position of unity, requiring that further acts of differentiation take place. It is this need to keep differentiating against a countervailing pull of similitude that gives New Guinea cultures their forward momentum. Unity and disjuncture become twin moments in the ongoing flow or elicitation of social life. The taboo conditions I describe below are noteworthy in that they capture both of these processes simultaneously. On the one hand, what they differentiate is a moment of unchecked similarity, and on the other they become a venue through which such differentiation is carried out. Through an examination of Kamea food interdictions, I ask what it might mean to our understanding of social life were bodily integrity and autonomy not taken as the underlying basis of personhood.

EATING FOR OTHERS

Food is an important nexus of cultural value for Kamea. Its consumption furnishes an interested observer with an analytic springboard into indigenous views of the human body. In chapter 2, we saw how prolonged feeding behavior in the form of consuming bridewealth payments gradually elicits a woman's reproductive capabilities. This process of engendering female agency runs counter to most interpretations of reciprocity. Contemporary anthropological theory holds that the sharing of food creates a state of "sameness" among people and is thus instrumental in the creation of kin ties (see Carsten 1995, 1997; A. Strathern 1973; Weismantel 1995). Yet, as we have already seen with Kamea, the sharing of food is equally important in eliciting salient social differences. And we shall see that if feeding is tied to the creation of social distinctions, taboo among Kamea often speaks to similarity rather than difference.[4]

I first became aware of the significance of Kamea food taboos a few weeks after taking up permanent residence at Titamnga. One morning, not long after I had arrived in the field, I was invited to join a small group of women and children to hunt rats (*mataka*) along the forest edge. After several hours of work, we had managed to capture a few of the rodents, and then beat a hasty retreat back to our homes to avoid the late afternoon rains. Later that evening, Kokoban, one of the women in our party, dropped by my house and launched into a detailed account of how she had distributed her share of the spoils to kinsmen. Everyone in her immediate family, it seemed, had received at least some portion of the game, although I noticed that she failed to mention her younger brother Drinda as having eaten any. When I asked Kokoban about it, she explained that rats were taboo to him. If Drinda's mother were dead, the boy could eat them without fear, but for the time being, at least, they were forbidden until he was initiated.

Rats, as it turned out, were not the only item prohibited to young boys. As I pursued the topic of food taboos in greater detail, I soon learned that boys were also barred from eating a number of other mammalian species, including several varieties of marsupials and bats. Many of these animals live on the ground, others in trees, and they are indigenously recognized as having highly variable habits and appearances. What each of these different species has in common with one another is that they are all said to emit a particularly strong odor when cooked.[5] While smell is a highly salient quality of experience for Kamea, my concern in what follows lies less with analyzing the semantic content of these prohibitions than it does with explicating the specific form that they take.[6] As we shall see, meaningfulness for Kamea resides less in a system of positively defined concepts than it does in the cracks and crevices that exist between them.

At the time it struck me as rather ironic that the more I struggled to learn about rats and boys, the more my friends at Titamnga wanted to tell me about the life cycle of women. Unlike a boy who labors under an onerous array of dietary restrictions, Kamea girls are free to eat virtually anything they please. Indeed, some women will enjoy a taboo-free existence until the day they die. The determining factor is the sex of any offspring they may carry. Should a woman give birth to a child of the same sex as herself, her life will continue pretty much as before; both she and her daughter are free to consume all varieties of "smelly" game. If she carries a son, by contrast, her eating habits will eventually change. From the time that her son has his nose pierced during the first phase of initiation, all of the food items that were taboo to him now become taboo to his mother as well. When both stages of the initiation sequence have been completed (discussed later in this chapter), the boy can begin to eat these items for the first time in his life, but they will remain taboo to his mother until her dying day. Should a woman violate these taboos, it is said that her "backside would break" and she would be forced to remain in the house, lonely and sapped of her energy. A woman so afflicted eventually dies of a wasting disease.

Thus far, the taboo conditions I have been discussing appear to fall quite neatly within the scope of orthodox anthropological theory. A mother and son never eat precisely the same foods at the same time—what is permitted to one is taboo for the other.[7] As Mary Douglas (1966) told us decades ago, food taboos are acts of separation that define discrete identities through a system of negative differentiation (i.e., individual *a* is whatever individual *b* is not). The situation at Titamnga, however, proved to be somewhat more complex. We have seen why a mother abstains from eating certain varieties of game; she follows these taboos to protect her health and physical well-being. But it remains to be seen why a son, prior to being initiated, should do the same.

Although a few Kamea told me that boys avoid these items because their consumption may lead to a scabies-like condition, this was seen as a relatively minor consequence; certainly a small price to pay for a tasty mouthful of much-prized game. Instead, the principal reason given for eschewing these items is that were a boy to eat them, they would have an adverse effect *not on himself, but on his mother.* As one boy explained: "If I ate these things, my mother wouldn't be able to leave the house, she couldn't go to her garden, collect firewood, or care for livestock. She would sit in the house all day long where she would eventually die."

For Kamea, then, what the initial taboo conditions mark is not a *singular* but rather, a *conflated* identity. In an insightful paper that deals with the Malagasy-speakers of Mayotte, Lambek (1992) discusses one of the more common features of taboo: "a taboo clearly differentiates between those who practice it and those who need not" (249). For Kamea, as we have seen, this is not necessarily the case. The taboo conditions that operate between a mother and son do not simply mark the two as distinct, but also establish their *essential sameness.* Taken together, they present an image of a singular body wherein the eating habits of one party directly affect the health of the other. But the fact that only half of this dual entity actually follows the taboo also anticipates the differentiation that will take place through the men's cult and its associated rituals.

Prior to being initiated, a young boy is said to be "female-like" (cf. Godelier 1986: 31; Herdt 1981: 203–54, 1987: 101–69).[8] Indeed, Kamea boys continue to wear the style of grass skirt that is characteristic of older women. It is only through participation in the men's cult that a masculinized form of being is created. At initiation, the previously conjoined identity of the mother and son is disarticulated, thereby creating the son's capacity to be a gendered agent (cf. M. Strathern 1993). The taboo conditions emphasize both phases of this process, and shifting them becomes a means through which this state of disjuncture is achieved.

I have described the intersection of cult life and taboos in terms of their capacity to call forth different types of culturally recognized identities. It should be noted, however, that Kamea ideas surrounding the efficacy of initiation translate rather awkwardly into the categories of Western social science. While Western-trained anthropologists are apt to see initiation as being constitutive of gender, Kamea are more apt to view it as a process of "decontainment," a process that carries with it a decidedly temporal dimension.

As I have demonstrated in previous chapters, Kamea kinship is bilateral. Social relationships in this system are organized along two planes: the lineal (male and cross-generational) and the lateral (female and same generation). A continuity with past generations is seen to adhere in the male line and becomes the basis through which claims to property are activated. Land, paternal names, and modes of ritual competence are all transmitted through men, typically from a father to his son. Siblingship, on the other hand, derives exclusively from women and is spoken of in terms of an idiom of "one-bloodedness." Although women are not "one-blood" with the children they bear, they bestow upon their offspring a horizontal mode of relating. Neither of these relational modes is based on a genealogical framework. This has important implications in terms of how Kamea sociality unfolds through time. It also significantly influences the degree to which Kamea social life can be grasped through existing theoretical models in the social sciences.

It has become commonplace in the anthropological literature to argue that male initiation detaches a boy from his mother and allows him to form a counteridentification with the community of men. Nearly half a century ago, Burton and Whiting (1961) set forth the idea that initiation turns maternally attached boys into psychological beings that identify themselves as men.[9] This feat is often accomplished, or so we are told, by manipulating bodily substances. As Herdt has written of the neighboring Sambia: "It is a matter of urgent concern that the mother's contaminated blood be removed from boys; otherwise male biological development is impeded" (Herdt 1981: 226, cf. J. Weiner 1982). As we have already seen, substance for Kamea *separates,* rather than *conjoins,* persons in proximate generations. Removing the "female contribution" that went into conception makes little sense when such elements are not understood to be imparted in the first place. This is not to say that the relationship between a Kamea woman and her child is not embodied; rather, the nature of the connection is not based on the notion of "shared substance." Instead, the bond between a woman and her child is based on that act which brought the relationship into being in the first place—the very act of "containment" itself (see chapter 2 for further details).

Until he is initiated, a boy is, in a sense, still "contained."[10] During the parturition process what the mother eats directly affects the health of her child. Pregnant women are enjoined not to eat either sago or watercress because both have deep root systems that will fasten the child to the uterus and prevent an easy birth. The taboo system marks an extension of this process in that the eating habits of one directly affect the well-being of the other. It is only through initiation that a woman's son is finally "decontained." He and his mother are no longer joined as a single entity and the consumptive patterns of each no longer affect the other. He who was once one with the mother is now free to enter into a relationship; the son can take a wife, form a household, and initiate a family of his own.

In the next section, I examine the men's cult in greater detail and the actual means through which this separation is achieved.

KAMEA INITIATION

Every four to six years, Kamea sponsor an initiation ceremony (*apa*) that is divided into two distinct ritual phases. A boy's entry into the men's cult begins when he is ten or eleven years old and is brought to the bush to have his nose pierced by an older initiated man. Along with his age mates, he will remain in the cult house for a number of weeks, until the sores have healed, after which time initiates are permitted to return to the house of their parents. Several years later, boys will be fed the fruit of the pandanus tree (marita) in a secret cult ceremony that marks the final grade of the initiation sequence. Passage through both these stages is marked by the disclosure of secret narratives, including detailed knowledge of how to conduct oneself in matters pertaining to warfare and women (cf. Blackwood 1978).[11]

Over the course of my research, I had the opportunity to witness the second stage of initiation, the marita ceremony, which was performed in March and April 1990. As will become evident in what follows, sometimes during this phase men and women share center stage; at other times, their activities are kept strictly separate. When their activities were separate, my observations were based on what took place among women.[12] My understanding of what takes place within the men's cult house, then, is based on what I could glean from male interviewees. The same holds true of first-stage initiation, which I never had the opportunity to witness directly. Here too, I rely on the verbal accounts of men and women.

When an initiation is about to get underway, a men's cult house (*hewa iya anga*) will be built in a secluded forest clearing, far from the supposedly watchful eyes of women and children. It is here that boys will be brought to have their nasal septum pierced during the opening phase of the initiation sequence, and here that they will remain to convalesce, a process that takes from between six to eight weeks to complete.

Prior to the influence of government and missionary officials (circa 1965), it was mandatory for all boys to be initiated. Through the activities of the men's cult, boys were given the necessary strength (*yannganga*) to carry out a range of male pursuits, including hunting game, clearing gardens, and climbing pandanus trees. The men's cult also instilled within boys an aggressive temperament, a trait held in particularly high esteem prior to pacification. As one man explained: "Never mind that you see everyone around you die. You cannot be afraid. You must kill people back. We strengthened boys so that they could stand up and fight." Until recently, initiation was also a necessary precursor to marriage. It was inconceivable for an uninitiated youth to take a wife. The men's cult taught boys how to behave in the presence of women and how to avoid being contaminated by the polluting and sexual substances of their brides-to-be.[13] Initiates were also instructed in the use of the *yangwa* tree as a means of replacing semen that has been lost through sexual intercourse (see chapter 1).

Within the contemporary context, it is up to the boy himself to decide whether or not to participate in the ceremonies.[14] On the basis of my research, I estimate that approximately 20 to 30 percent of all boys will choose to undergo full initiation rites, meaning that they will sport a pierced nasal septum as an adult. It is important to note, however, that this does not give an entirely accurate picture of where things stand. A truncated form of initiation is emerging wherein boys are taught all of the secrets of the men's cult and are shown the bullroarers (*mautwa*) but refrain from having their nose pierced (see Bamford 2006). This is done, I was told, in order to hide the men's cult from local missionaries, who have been relentless in their campaign to put an end to initiation practices since they first began to work in the region during the early 1960s. Because these men do not embody any visible sign of their changed status, it is difficult to know how many men have participated in this abbreviated system of rites.

Today, any boy who wishes to participate in an initiation sequence is free to do so. The ceremony that I witnessed in March and April 1990 (to be described below) exhibited marked variability, both with respect to the age of the initiates themselves and in terms of the villages from which they hailed.

The youngest boy in attendance was approximately ten years of age while the oldest was somewhere in his mid to late twenties. Most of the participants were drawn from within a fifteen-mile radius, although I was told that persons who lived further afield could have participated in the rites if they so desired. Prior to the arrival of colonial agents, this system exhibited far less flexibility; all boys between the ages of nine and twelve years were initiated. Age mates were generally drawn from the same pool of persons who might be called upon to assist one another in times of war. This fact fits well with one of the intended aims of the men's cult, the promotion of an aggressive warriorhood.

While boys are secluded in the men's house, they remain under the strict supervision of older, initiated men. In most cases, a mother's brother will take it upon himself to oversee the ritual fate of his sister's son. In addition to having their nasal septum pierced, young boys are beaten repeatedly with sticks. This is believed to promote qualities of strength and endurance and has the added benefit of teaching boys how to fight. Initiates are also shown the bullroarers for the first time and are taught how to turn them in order to produce a high-pitched whine. Notwithstanding these activities, the bulk of their time in seclusion is spent on long and arduous hunts that range over wide tracts of forest. Novices are schooled in the art of setting traps and tracking animals, and in the proficient use of the bow and arrow. Older men tell boys that they cannot be afraid of the dark because the night is an excellent time to hunt game (<u>kapul</u>). Through it all, women are strictly forbidden from approaching the cult house. According to one man: "If women see what is going on, the noses of initiates will break. Later, the woman herself would become sick. She would lose all of her strength and die."

What I have described thus far sounds similar in all but its details to what has been written of male initiation elsewhere in the highlands of Papua New Guinea. Through the rituals of the men's cult, what is seen as an androgynous or effeminate being is transformed into a masculine form. New Guinea ethnographers have always been impressed by the fact that Melanesians appear to make considerable use of gender imagery. In many accounts, male and female appear as polarized beings who are not only different but anathema to one another. Their distinct sexual substances and physiological processes are seen to have a negative effect on members of the opposite sex. The existence of a gender dichotomy is also reflected in a strict division of labor, wherein males and females are not only assigned to different productive domains but are believed to be incapable of carrying out the work of the other.

Most interpretations of male initiation in Melanesia have taken the existence of this dichotomy as their baseline and have attempted to document how it comes to be reproduced generation after generation. The men's cult is seen as an important vehicle through which a male identity is stamped on the body of a boy. But more than the creation of a gendered being is at stake. In many accounts, the rituals of the men's cult emerge as a political weapon through which men instantiate and perpetuate their political domination over the world of women (see, in particular, Godelier 1986). Many anthropologists have claimed that through initiation, men usurp the powers of women to give birth (Bonnemère 1996; Herdt 1981; Godelier 1986), thus asserting their autonomy and political ascendancy: "Initiation among the Baruya and Sambia is a protracted and painful process of embodiment which underlies the detachment of the boy from his mother and children. Indeed, the separation of sons from their mothers 'is at the heart of the issue' for what men do 'in the imaginary world' is to dispossess women of their creative powers and to transfer these to men" (Herdt 1996: 7). Initiation emerges as a war between the sexes: one that pits men against women and reproduces a system of unequal power relations.

As I shall demonstrate, Kamea initiation is not about reproducing a hierarchy of gendered states. They do not see the world as resting upon a ready-made system of social and biological distinctions in need of being perpetuated: it is the act of *constituting* those distinctions in the first place that is the aim of social action. The aim and intent of the initiation sequence reflects this shift in emphasis. Kamea women are intimately involved in the initiation of young boys; indeed, it is the mother-son relationship that stands at the heart of the initiation sequence. Men do not reproduce other men independently of women. Rather, both sexes engender the social world through their mutually elicitive actions (M. Strathern 1988, 1993; Wagner 1975, 1977). There are no exclusive camps that are capable of their own parthogenesis because salient distinctions are understood to be *produced* rather than *reproduced*. A boy is not initiated into the men's cult so much as he is detached from an encompassing female form. Women are not incidental to the process; they furnish the ground and the motivating force against which it takes place.

While the men's house is being raised during the opening phase of the initiation sequence, a second collective dwelling will be made for the mothers of the boys. During the entire time that their sons are in seclusion, the mothers will be confined to this second dwelling. Confined, the women are subject to a number of interdictions, particularly, those concerning the use of space. In the words of one man: "Mothers can't go to the bush, they can't go to their gardens, they can't collect sweet potatoes or search for firewood in the bush. These items will be brought to them in seclusion. The only thing mothers can do is sit in the house. If they leave it, the noses of their sons will break." The image we are presented with is that of a container who is tightly contained. It is only when boys are released from their own seclusion several weeks later that the women can quit their confinement. Throughout this time, the fate of a mother and son is completely intertwined.

It is during this phase of the initiation sequence that the dietary taboos discussed above undergo their first important shift. As noted earlier, while a boy is growing up, he is prohibited from eating a wide variety of "smelly" game. Once he completes the first stage of initiation, these food items become taboo to his mother as well. Thus, for a period of approximately one to two years (i.e., until the boy is fed marita), he and his mother are united by their joint adherence to these taboos. Should either the mother or the son consume one of the varieties of interdicted game, the effect would not be felt by the one who broke the taboo, but on the body of the other. After a boy has completed his final stage of initiation, he can eat these items without fear of ill effect. However, they will remain taboo to his mother for the entirety of her life. A woman who eats these items risks not only illness, but also premature death. It is significant that it is only after a boy has completed both stages of initiation that he and his mother achieve a measure of autonomy in their eating habits. Henceforth, a mother will follow these taboos to protect her own well-being. Initiation has succeeded in separating what was previously a conjoined social identity.

In addition to the interdiction against "smelly" game, other foods become taboo at initiation. A boy who is recovering from having his nasal septum pierced will refrain from eating pitpit, mushrooms, and marita, along with certain varieties of sugarcane. All of these items are considered "greasy" (wel); if they were consumed it is believed that the boy's nasal septum would remain "wet" and never heal. After he has been released from his seclusion, the boy is free to eat these items once again. It is important to note, however, a point that is once again indicative of the nature of the mother-son bond: that the mother of the initiate will also refrain from eating these items for precisely the same reason—were she to consume them her son's nose would "break." The mother and son operate as a singular body. But unlike the tabooed game, these vegetal items can be consumed again after the nose-piercing ceremony has been completed. Their evocation during the first stage of initiation further signifies the embodied nature of the mother-son relationship, and the fact that this relationship is being altered through the performance of these rites.

I noted earlier that rituals of male initiation have long been seen as an important vehicle of political domination. The ethnographic literature is replete with examples in which men's control over the ritual processes and sacred cult objects both validates and legitimizes an unequal balance of power between the sexes (Godelier 1986; Herdt 1981; Langness 1967; Read 1952). Within this context, it is interesting to note that one of the most important tasks assigned to Kamea women during their seclusion is to guard the bullroarers (*mautwa*), the preeminent symbol of male cult life throughout Melanesia. Just before mothers of boys enter their cult house in the bush, they will be given the bullroarers by men, along with instructions to watch over them carefully. The bullroarers are wrapped in leaves and pieces of bark cloth before being presented to women, leaving only the tip of one end exposed. According to men, women are not fully cognizant of what they are holding.[15] They know they have been given an important item of the men's cult, but they are ignorant of the fact that it is this which later cries out during the final phase of the first-stage initiation. As one man explained: "*Mautwa* is something for men only. When we give it to women, it is carefully hidden. Each woman will take care of one bullroarer. When they hear the *mautwa* cry, they will say, 'Something is calling out to our children now.' Women don't know that it is the *mautwa* that cries. They think it is an insect or something like that."[16]

Later, toward the end of the boys' confinement, the older men will retrieve the bullroarers from the women and bring them back to the men's house, where they will be shown to initiates for the first time. The process of recouping the *mautwa* parallels and anticipates the eventual "decontainment" of the boys themselves, which is finally achieved during the final stage of initiation (see below). What is male must first be contained within an encompassing female environment; its engenderment as a male form is brought about by a process of extraction.[17]

The marked separation of mothers from everyone else in the community has a second, related rationale. It prevents the woman from becoming pregnant while her son is undergoing first stage initiation. As one man explained:

When a boy is being initiated, his father cannot go and look at his wife. The father must live on his own. The mother must live on her own. When the boys' sore is finished, they can live together again in the same house. But while a boy is being initiated, his parents must be separated. It would not be good if the two engaged in sex and they began to work on a new child. The nose of the boy being initiated would break. Pus would come up on the sore of the child. His nose would be infected and it would eventually rot away. A mother must go to live on her own and a father must live on his own. That way, they won't desire each other too much. When the sore is finished, they can sleep together again.

If an important aim of the men's cult is to separate (that is, to "decontain") a boy from his mother, it makes sense that a concerted effort would be made not to evoke the woman's "containing" capacities at this time. To produce a second child would be to elicit an act of maternal encompassment at a time when every effort is being made to break a previous act of containment.

When the sores of the initiates have completely healed, a celebration takes place, during which both initiates and their mothers are permitted to leave their respective cult houses. At this time a small feast is held and the accumulated game is presented to maternal kin. It is significant that as boys and their mothers quit their confinement a mixture of yellow clay and dried leaves is rubbed on the bodies of each. This same mixture is spread on a mother and her newborn infant when they leave the birth house and prepare to enter the regular family domicile. This point calls to mind the long noted association between the symbolism of initiation and the process of childbirth (Van Gennep 1960). Yet, here it is not men who take on female powers and channel them to their own political ends; instead, both sexes are equally involved in creating a new gendered form. As Marilyn Strathern has noted of childbirth more generally, "Women's giving birth as an act cannot be taken over by men; men can only intervene as causing the action. . . . A mother is coerced into becoming a mother" (1988: 332). For Kamea, it is perhaps appropriate to suggest that men "coerce" women into "giving birth" to children that they have contained for nine to twelve years.

Approximately one to two years after the first stage of initiation has been completed, boys undergo the final phase of the initiation sequence, which involves being fed the fruit of the red pandanus tree. It is to an account of this rite that I now turn my attention.

In March and April 1990, I had the opportunity to witness the second stage of Kamea male initiation. The rites were held in a small clearing approximately three hours' walk due west of Titamnga. Fifteen initiates and several hundred of their friends and relatives took part in the ceremony.

The final stage of initiation lasts from one to two months. Throughout this time, rites are held on a nightly basis, beginning at dusk and continuing until dawn the next day. Although initiates are expected to attend each evening's festivities, those people who participate as dancers and audience members come and go as their schedule permits. The mothers and fathers of initiates generally attend the ceremonies every evening, while less distant kin drop by the dance ground only intermittently.

When I arrived on the scene, the second stage of initiation was already well underway. I was able to witness the last few days of ritual activity leading up to (and including) the point at which boys are fed the juice of the pandanus fruit. My consultants told me that what I saw during this period of time was representative of what takes place during the second-stage initiation; it is only on the final night (to be described below) that the ritual sequence changes in both form and content.

In order to give a fuller account of what takes place during these rites, I will weave back and forth in the description that follows between a narrative account based on field notes and a more analytic discussion intended to situate various aspects of the ethnography. These vignettes also help to situate me as an ethnographer and the Kamea reactions to my presence.

ENGENDERING MALE AGENCY: THE FINAL PHASE

We arrive at Hawabango at about six-thirty in the evening, after a three-hour walk in the pouring rain. I am accompanied by Bipahu, Cajo, Sekikawa, Nonjos, Medi, Sasun, and Hipil.

Hawabango is a small Catholic mission station situated five miles to the west of Titamnga. Like Kaintiba (the larger government outpost to the south), it sports a few trade stores and is serviced on an ad hoc basis by flights from the coast (i.e., Kerema) and the Morobe interior. The initiation site is still one hour's walk away. Iva, a resident of Hawabango, tells me that the onset of the heavy rains will have delayed the evening's activities by several hours. We decide to rest and to dry ourselves in a small thatch house that flanks the eastern corner of the station.

Immediately after we arrive at Hawabango, Medi and Sekikawa set off by themselves to ready their decorations (bilas) for the upcoming ceremony. The main item of ritual attire worn by men during the final initiation grade is an elaborate headdress, which is made of bird of paradise and cockatoo feathers. The headdress is approximately three feet in length and is shaped like the letter *A,* to which is attached an elongated center appendage. This centerpiece, which is made of cane and cockatoo feathers, bobs up and down in keeping with the rhythm of the men's dance. Medi and Sekikawa spend the next several hours straightening their headdress in an effort to ensure that it will have the desired effect when they dance.

By ten o'clock, the rain has subsided and we set off for the initiation site. The blackness of the night is filled with lit <u>mambu</u> torches carried by men and women, who, like ourselves, have taken advantage of the break in the rains to make their way to the dance ground clearing. From a distance, these lights appear to be tiny fireflies in the night sky, weaving their way up and down the sides of mountains, but progressing in the same general direction. We walk for about forty or fifty minutes and finally reach a place where a small footpath bends off to the left. From here, it spirals off in a downward direction, where it is eventually lost in the blackness of the night. I can neither see nor hear much in the way of activity down below. Medi calls out to inquire whether the <u>singsing</u> (ceremony) is still on. We are told that it is, and begin our descent into the awaiting darkness.

The initiation site consists of a large cleared space, roughly fifty feet in diameter, which is now packed with mud because of the extensive rain. A small fire has been lit near the center, around which a group of men and women huddle in an attempt to cast off the chill of the mountain air. Several makeshift shelters line the perimeter of the dance ground. I am told that these can be used as a place to sleep if one tires before the all-night proceedings have reached their conclusion. I am aware that my companions think I will have drifted off long before dawn arrives.

As more and more people enter the clearing, the fire at the center of the dance ground is extinguished. Girenia, Ip'i, and a few other women from Titamnga with whom I am acquainted take hold of my arm and attempt to drag me to the outer reaches of the dance ground. I protest that I want to see what is happening in the middle, but they are adamant that my place is with them along the outer edges of the clearing. I allow myself to be led away. Girenia grabs hold of my hand and begins to escort me around the clearing. At first, I think that she is looking for someone she knows, but she continues to lead me in a counterclockwise direction. An old woman on my right places a <u>mambu</u> torch in my hand. I take hold of it and carry it with me as I walk. Sixty people or so are now in attendance. More enter the clearing with each passing minute.

As I become more familiar with my surroundings, I begin to notice that whenever a newcomer enters the dance ground, he or she will file in with the others and begin to walk or dance around the clearing in a counter-clockwise direction. When people tire of this activity, they go to watch the proceedings from the sidelines. Here, they rest, enjoy a smoke, and engage in casual conversation with friends or relatives. The atmosphere is informal and no one seems to be the least bit disturbed by my presence.

As the night progresses, I focus more on the staging of the ceremony. The initiates form a small, stationary group at the center of the dance ground. Their fathers and other male kinsmen surround them, and at the outer fringe of the dance ground move the women. Several of the older women carry large staffs (*cot'wa*) in their hands. Girenia explains that these women are the mothers of boys. The staff they carry reminds me of a digging stick, only it is longer and made from an entirely different variety of wood. Men and women each have their own unique styles of dance. Men adopt a short and bouncing gait as they move, which causes the third appendage of the headdress to snap up and down. Women possess their own distinctive style. They will move forward for a step or two, turn inward in a half-circle, and then move forward for a step or two again. The younger girls laugh and tease one another as they try to imitate the steps of the older women.

The bulk of the evening passes this way, with men and women moving about the dance ground in the manner described above. At some point during the night—I seem to have missed the exact moment at which it occurred—a large wooden post is erected at the center of the dance ground. I make several inquiries about the significance of the post and I am told that it helps to strengthen the efficacy of the singsing. I notice that initiates now stand with their faces turned toward the post and their hands outstretched as if supporting it. As the night progresses, a few people retire to the sidelines. The rest continue their circuit around the dance ground.

As dawn approaches, a new component is introduced into the dances of men and women. The initiates and older men who have joined them at the center of the dance ground join hands and begin to jump up and down. As they do so, they emit a high-pitched cry. The mothers of the boys do the same thing from their position at the periphery of the clearing. This continues off and on for the next twenty to thirty minutes. An old man from one of the neighboring villages with which I am not familiar explains to me that the ceremonies are about to come to a close for the night—people are celebrating (hamamas) the completion of another day of initiation.

As the sun rises, people leave the dance ground one by one and slowly make their way home. There, they will sleep, eat, and carry out whatever

Figure 6. Women dance during the second stage of initiation. Photo by Sandra Bamford.

other tasks are necessary before returning to the dance ground the following evening.

Despite the changes in the personnel attending these rites, second-stage initiation follows the same trajectory night after night. Boys always occupy the center of the dance ground, surrounded by men, who are in turn encompassed by women. The evening passes with everyone moving around in a circle. When first light breaks, everyone disbands and returns home. It is not until the final night of initiation that significant changes are introduced into the ritual sequence.

An atmosphere of excitement and tension fills the air. Tonight the boys are to be given <u>marita</u>. I sit on the verandah of my house at Titamnga and wait for my companions to ready themselves. In contrast to all other nights, everyone is paying far closer attention to his or her dress and <u>bilas</u> (decoration). Men and women sport newly made <u>pulpuls</u> (grass skirts)—the kind that comes to a sharp point in the front, rather than the usual style, worn for day-to-day use, which is cut straight across. Noniha approaches and sits down to wait with me. She is wearing several strings of shell ornaments (*nuwa*) about her neck and has put on enough *ituka* (arm bands) to almost completely cover the flesh of her upper arms. Several strings of <u>mambu</u> (bamboo) are worn across her chest, their yellow color furnishing a dramatic contrast to the rest of her costume. She chatters happily and munches on some taro while we wait for the rest of our party.

We arrive at the dance ground at about eleven o'clock. Already it is filled with people, far more than have been in attendance on previous evenings. Those of us who just arrived from Titamnga file in with the others and begin our now familiar circuit about the dance ground. As the people move, they sing songs in the <u>tok ples</u> (the Kapau language). I ask Vadid [a man from the village of Wembango], who walks beside me, what they are singing about. He responds: "We sing of plants, trees, fruits, stones, mountains, rivers, and the forest. We comment on the beauty of the place; of the rain, of how the water flows, and of where the sun goes where it sets at the end of the day."

It is entirely appropriate that the songs sung during the second stage of initiation draw upon richly evocative images of the physical environment. Before he is initiated, a boy is a being of undetermined gender who is strongly attached to (if not indistinguishable) from his mother. Initiation songs foreshadow the world that the youth will now be entering as a consequence of his participation in the men's cult. As we saw in chapter 1, male sociality is objectified in the relationships that men form with the nonhuman world. The fact that men draw upon this world to supplement their own stores of semen further strengthens the conceptual connection between maleness, intergenerational relationships, and the land.

After a boy has been initiated, he will enter a world where he can act directly on the nonhuman world himself. Up until this point in his life, he has relied exclusively upon his parents to meet all of his subsistence needs. Young boys are expected to do little in the way of contributing to the family economy, spending most of their days roaming the bush in the leisurely company of friends. From the time that he has completed the <u>marita</u> ceremony, an initiated youth (*hewa*) will begin to make gardens on his own. His father will assign him a tract of land that he will begin to clear in anticipation of supporting a wife and family. It is no accident that the songs sung during the final stage of initiation focus on naming different features in the physical environment, particularly trees, mountains, streams, and plots of ground. To sing these songs is to give concrete reference to that domain of activity that a youth will now enter.

I ask Vadid, who is fluent in both Neo-Melanesian and Kapau, to speak to the mothers of boys on my behalf. I am interested in knowing how the women perceive their own role in initiation. How does their presence enhance the efficacy of the rites? The women smile mysteriously at him and respond with what must strike them as an obvious adage: he is a man—it is not his place to know what happens with women at this time.

As noted in the introduction, Kamea understand knowledge to have corporeal effects. It is not so much a question of each sex carving out an exclusive sphere of influence, so much as secrecy is seen as a health-related issue. In the wrong hands, knowledge is a toxin—a poison. Should either sex learn what is appropriate only for the other to know, sickness or death is the inevitable outcome. To keep a secret, then, is to demonstrate concern for the welfare of another.

At the top of a hill, roughly seventy-five feet from the dance ground, stands a medium-sized cult house where boys will be fed <u>marita</u>. I do not remember seeing this structure on any of the previous nights, and when I ask my companions about it, they explain that it was built only recently in anticipation of the final phase of initiation. The cult house will be razed in the morning when the second stage of initiation has been completed.

At about two o'clock in the morning, the ceremony takes an abrupt turn. The men at the center of the dance ground begin to jump up and down. They take hold of the initiates and race toward the cult house on the hill, but before reaching their destination, they turn on their heels and reenter the dance ground. This is repeated several more times. On the fifth charge, the older men disappear with the initiates into the cult house. Those of us who remain down below continue with our circular pilgrimage. I ask where the initiates have gone and receive two different answers: (1) they have gone to cook different varieties of game; and (2) they are being given <u>marita</u> by older, initiated men.

One of the most carefully guarded secrets of initiation concerns what actually takes place in this cult house on the hill. The official and public account (i.e., that which is known to both women and children) holds that youths are fed <u>marita</u>—the bright red fruit of the pandanus tree. Through this act of consumption, they are finally released from all of their dietary taboos, including the taboo against "smelly" game, and the second stage of initiation begins to draw to a close. What actually does take place in the men's cult house is notably different from the official canon. Instead of being given pandanus fruit to eat, the juice of this fruit is rubbed on the face and bodies of initiates. Boys will be fed <u>marita</u> only later—after they have first been thoroughly coated with this substance from head to toe.

To grasp the significance of this rite, it is necessary to understand something of the importance of pandanus fruit (<u>marita</u>) in Kamea thought. On several occasions, I was told by my friends at Titamnga that <u>marita</u> had "come up" (that is, originated) from the "blood of all men." We saw earlier (chapter 1) that in the Kamea myth of origins, all of humanity spilled forth from the trunk of a tree. After having undergone a series of important transformations, Akeanga's charred remains metamorphose as human beings who take shelter inside a tree. The two sisters make a series of incisions in the trunk and each time they do so, another collectivity of men is released. As the men tumble forth from their wooden environment, afterbirth spills all over the ground, pandanus fruit trees sprouting where it falls.

<u>Marita</u>, then, is explicitly associated with afterbirth in indigenous thought. It is significant that this substance must first be rubbed on the skin of boys before their own bodies can come to contain it through an act of ingestion. An encompassing female form must first contain that which is "male"; it is only later that the capacity to act with a measure of autonomy is achieved.

The light from a fire illuminates the cult house on the hill. From down below, the house appears to be an incandescent beacon, glowing brightly against the velvet blackness of the night. I long to be with the boys and men myself, as much to warm myself as to catch a glimpse of what is going on. Instead, I walk in a circle. Girenia bites down hard on the tough outer skin of some betel nut and wraps her bark cape more securely about her shoulders. A few children doze in the makeshift shelters along the sidelines. I fidget distractedly and wonder when the monotony of going around in circles will end.

At about four o'clock in the morning, a number of <u>mambu</u> (bamboo) torches appear on the hillside. One of the men who has chosen to remain down below instructs me to keep well out of the way—otherwise I am likely to get hurt in the upcoming commotion. I retire to the outer reaches of the dance ground along with other women.

The cry of bamboo flutes fills the air. The initiates, along with several older men, storm noisily into the dance ground. Their arms are linked at the elbows and they are crouched down low to the ground in a squatting position. As they begin to fill the clearing, they fan out to form an ever-widening circle. As the circle expands, they begin to bash into the spectators on the sidelines. I am glad that I was warned to stand a respectable distance from the goings-on. Almost as abruptly as it began, this phase of activity ends. Everyone resumes their accustomed positions from the previous nights: boys in the middle, women along the outer edges. Our journey around the circle begins anew.

At dawn, several things take place. Two large posts are carried into the clearing and are placed horizontally on the ground, forming two parallel lines facing the cult house. The initiates are instructed to sit on these posts. The mothers and sisters of boys stand behind their sons and brothers. The rest of us watch from the outer parameter of the dance ground. The atmosphere has suddenly become tense. A number of the older men have returned to the cult house on the hill, where they begin to sound the bamboo flutes again. Suddenly, two men rush into the dance ground. They run a circle around the boys who are seated and then tear back to the cult house again. The youngest of the initiates begins to cry. Ten minutes later, this sequence of activities is repeated. It occurs for a third, forth, and fifth time over the next half-hour. Each time the men charge into the dance ground, the mothers of the boys emit a piercing cry and begin to jump up and down. The overall effect is quite dramatic. Those of us who watch from the sidelines are mesmerized.

Suddenly, everything comes to a halt. Nearly an hour goes by with no further sign of activity. The initiates begin to look bored and one boy dozes, perched on his seat. Loma, who is seated beside me, complains about the heat. Now that the sun has risen, it is beginning to get quite warm. One of the mothers of the initiates leaves the dance ground and disappears along a footpath that leads off into the surrounding forest. She returns a short time later with a wet cloth and proceeds to mop the face and upper body of her son. Several other women follow suit. We seem to be waiting for something, but I do not know what.

Mambu flutes are, once again, heard from the hillside. As before, two men race into the dance ground. This time, however, they do not return immediately to the men's house: they bend close to the initiates and seem to be giving them some kind of instruction. More men pour into the dance ground. The initiates rise and begin to run with the men to the cult house on the hill. Almost as abruptly, however, they stop and reenter the dance ground. This pattern of activity is repeated several times. The howls of the youngest initiate have reached a frantic pitch.

Several men bearing steel knives race into the clearing. They begin to run clockwise around the boys. The mothers of the boys circle in the opposite direction, forming a barrier that contains the ritual action. This continues for several minutes. Then the men, accompanied by initiates, make a dash toward the cult house on the hill. A number of the women at-

Figure 7. Boys wait to be taken to the cult house for the final phase of initiation. Photo by Sandra Bamford.

tempt to follow their sons, but they are chased back by knife-wielding men. For the first time, I notice that a post has once again been erected at the center of the dance ground. Five women, coated from head to toe in mud, are supporting it.

I noted earlier that shortly after a woman gives birth, both her body and that of her baby are rubbed with mud. Kamea assert that this offers them a measure of protection against would-be sorcerers who are known to prey upon persons in a weakened condition.

The symbolism of the final phases of the initiation sequence emphasize a process of extraction whereby boys are detached from the containing influence of their mothers. Throughout the bulk of these rites, women—in particular, the mothers of boys—form an outer wall by positioning themselves along the periphery of the dance ground. Women collectively envelop the bodies of others, just as a mother envelops the body of her child throughout gestation. At the penultimate moment of the initiation sequence, the containment of women is finally and irrevocably broken. Initiates in the company of older men break through the line of women, who continue to dance in a circle. The staging of the rite mirrors its intended consequences.

Figure 8. A man sports a ceremonial headdress during
the final phase of initiation. Photo by Sandra Bamford.

DISCREPANT SUBJECTS

This chapter opened with a discussion of emerging fetal rights legislation
in the West. What recent court cases in Europe and North America reveal
is a deepening emphasis on bodily integrity—defined in biological terms—
as the underlying basis upon which legal personhood is assigned. This conceptual framework not only construes mother and child as separate individuals, but also as beings with potentially competing rights.

Feminist legal scholars have been quick to launch an assault against this conceptual framework. In particular, they have argued that while the concept of the free, autonomous, and equal person is fundamental to modern liberal legal systems, it also reflects a patriarchial and masculinist view of the world. The law's conception of a person with discrete bodily boundaries, they contend, misrepresents the embodied realities of women whose deviance from this norm is most apparent when they are pregnant. As Savell (2002) puts it:

> I have been arguing that the pregnant woman who carries a viable foetus does not fit law's conception of the legal person as a bounded unitary self. Her deviation from the embodied state of the legal person is located in the various strategies which law adopts to create a boundary between her and her foetus. When this occurs, control of her body/self may be ceded to other authorities who police this boundary by, for example, determining the conditions under which she might lawfully terminate her pregnancy, or make decisions concerning the delivery of her child, or whether she should be incarcerated to ensure that she does not pollute the maternal environment. (65)

Critics of fetal rights have increasingly been calling for a new vision of the legal subject, one whose identity is intrinsically mutable, defined in terms of an "openness" to others. Karpin, for example, suggests viewing the maternal body as a "nexus of relations." This framework would urge the law to see the mother and fetus (and later the child) as "intimately connected selves in a physical or material sense as well in a social and political sense" (Karpin, quoted in Savell 2002: 66). In contrast to the legal approaches discussed earlier in this chapter, such an account would "resist conceiving the foetus and woman as separable and perhaps oppositional and would instead explore the connective aspects of the relationship" (Savell 2002; cf. Naffine 2002).

What feminist scholars, then, have attempted to bring to light in our system is the diminution of the human being as a legal subject. In particular, they have attempted to show that our current model of legal personhood does a great disservice to women. Critics have also argued for a new—not to mention, radically different—vision of selfhood, one that would far more closely approximate Kamea understandings of the world, insomuch as it would theorize a personhood that has the possibility of being plural, rather than exclusively autonomous in nature (Naffine 2002: 89). Emmanuel Lévinas has described a relational self that comes into being through an openness to the other. Jennifer Nedelsky has similarly theorized a legal person whose freedoms stem from "enabling relations" with others (cited in Naffine 2002: 89).

Rather than taking strides in this direction, courts in Europe and North America increasingly appear to be taking up an opposite logic. Consider, for example, the issue of whether or not frozen embryos have protectable rights under Anglo law. This question made headline news in Melbourne, Australia, during the 1980s. In June 1981, Mario and Elsa Rios were killed in an airplane crash. They were owners of a million-dollar estate and had no living heirs. The question facing the courts was whether or not the IVF embryos they had stored at Queen Victoria Medical Center prior to their death had any heritable rights to their estate. Were the fertilized eggs property? Were they persons? If they were persons, who was responsible for their support? Was there any parental relationship? Should the state arrange for a surrogate mother (Wilson 1984)? Motivating these questions, and others like them, is an attempt to reconcile social rights and obligations with a vision of personhood that eventuates from biological reproduction.

Kamea cultural logic rests upon a radically different set of understandings. I have argued that Kamea rites of initiation separate what is otherwise a conjoined entity—the inherent singularity of a mother and son. Up until the time that he is initiated, a boy's identity is fully enmeshed with the identity of his mother. The nature of this attachment is strikingly revealed in the consequences of their respective eating behaviors; that is, by the capacity of each to affect the corporeal well-being of the other. Bodies are not perceived by Kamea in exclusively individual or relational terms, but can exist as both singular and composite states (M. Strathern 1988). One of the key ways in which bodies are brought together (as in sexual relations) and separated (as through initiation) is by being gendered, a process that eventuates, in part, through the performance of the initiation-taboo complex.

To be male in the world of Kamea is to be "decontained"; to be female is to possess the capacity to act as a "container" oneself. This distinction helps to cast light on a significant contrast between Kamea and what has been written of their immediate Anga neighbors to the north. Many Anga groups initiate women as well as men (see Godelier 1986: 40–51), but Kamea perceive such a practice to be erroneous. Whenever I questioned men and women about their views concerning the feasibility of initiating women, they found the entire idea to be quite unimaginable. In terms of the argument being advanced here, it can be seen that such a practice would make little sense. A female child will one day be a "container" herself and Kamea hope to encourage, rather than deter, the development of this capacity in girls. Women are the unmarked category here; it is maleness that requires additional explanation. Cultural understandings of gender and kinship are inseparable from the ritual system.

The last thirty years have witnessed many important changes in the initiation-taboo complex. Yet, significantly, instead of challenging those ideas that underlie the ceremonial sequence they have buttressed what is seen as its primary aim—"decontainment." Within the contemporary setting, if the mother of a boy is no longer living, he need not adopt the dietary restrictions that prohibit the consumption of "smelly" game. Safeguarding the health of his mother has ceased to be an issue and he can eat the tabooed items without fear of ill effect. By his mother's death, the son has already been "decontained."

While living at Titamnga, I had the opportunity to befriend a young boy whose mother had died several years earlier. Bipahu was approximately ten years old and had yet to be initiated. However, unlike most other boys his age, he was well versed in the goings-on of the men's cult and was given to turning the bullroarers whenever the opportunity presented itself. When I expressed surprise that a child such as Bipahu had any knowledge of the bullroarers, given that they were normally hidden from women and children, I was told that motherless boys are not barred from secret male proceedings, but are free to take on many of the rights of older men without having to undergo initiation themselves.

Kamea initiations are intimately bound up with the ways in which they see bodies and the connections between them. We have seen that the parent-child relationship is not imagined in terms of a tie of shared bodily substance. This carries with it a number of important implications. Much of the published literature on highland initiations takes the existence of a substance-based universe as its analytic baseline. Rituals such as bloodletting and insemination are understood to "remove" the maternal (i.e., female) part of the individual (see Herdt 1981: 203–54; Read 1952: 15; M. Strathern 1993: 47–48), replacing it with an exclusively masculine counterpart. Initiation, in other words, achieves its efficacy by acting directly upon the alchemy of bodily substances.

While this analytic model has proven to be highly useful in understanding the initiation systems of many Melanesians, Kamea fit rather awkwardly within the prevailing paradigm. Here, the tie between a woman and her child is not substance-based. Indeed, as we have seen, substance disconnects rather than conjoins persons in proximate generations. This is not to say that the connection between a woman and her children is not embodied; rather, it is based on that very act that brought the relationship into being in the first place: the act of containment. It is this containing influence that comes to be acted upon within the context of Kamea initiations. Both men and women are central to this process; the end result is not the reproduction of an enduring political hierarchy, but rather the ongoing creation of culturally meaningful social distinctions (Wagner 1974, 1975, 1977). Understanding women's participation in these rituals is more than mere "additive analysis," or a simple insertion of otherwise missing ethnographic facts. It leads to a fundamental change in perspective concerning the meaning of these rites and their consequences. What gets produced in this system is a shifting assemblage of identities and relationships, rather than a predetermined set of relationships that derive from heterosexual reproduction.

FOUR

(Im)Mortal Undertakings

On February 23, 1997, the world woke up to a technological innovation that "shook the foundations of Western biology and philosophy" (Silver 1997: 91). On that day, Ian Wilmut and his colleagues (1997) at Roslin Institute in Scotland announced the existence of Dolly—a six-month-old lamb who had been cloned from a single cell taken from the mammary gland of a Finn-Dorset ewe (Campbell et al. 1996; Wilmut, Campbell, and Tudge 2000; Wilmut et al. 2002). Although scientists had been cloning sheep and cattle from embryo donor cells for well over a decade (Pennisi and Williams 1997; Wills 1998), the birth of Dolly was nothing short of revolutionary. Unlike her predecessors, who had been cloned from cultured fetal cells, Dolly was the first mammal known to have been produced from an adult cell and one generation removed from the fertilization event that brought together the gametes of her genetic parents (Silver 1997: 100). In an interview conducted with science journalist Colin Tudge, Ian Wilmut— the chief scientist on the project—described the significance of his work in the following way: "Until the birth of Dolly, scientists were apt to declare that this or that procedure would be 'biologically impossible'—but now, that expression seems to have lost all meaning. In the 21st century and beyond, human ambition will be bound only by the laws of physics, the rules of logic, and our descendants' own sense of right and wrong. Truly, Dolly has taken us into the age of biological control" (Wilmut, Campbell, and Tudge 2000: 5).

The technique that made the cloning of Dolly (and now other animals) possible is a process known as somatic cell nuclear transfer (SCNT).[1] In this procedure, scientists extract the genetic material from a mature but unfertilized egg. They then inject the enucleated egg with the nucleus of another cell—a process known as nuclear transfer. The reconstituted egg is then activated with chemicals or an electrical current, which prompts it to grow and commence cell division (Cibelli et al. 2002).[2] Since nearly all of the hereditary material of a cell is contained in its nucleus, the renucleated egg and the organism into which it develops are genetically identical (except for the mitochondrial DNA) to the individual from which the transferred nucleus was derived (President's Council on Bioethics 2002: xxv). In the experiments carried out at Roslin Institute by Wilmut and his team, the reconstituted egg cells were then implanted into a foster mother, where they were allowed to develop—in Dolly's case successfully (Beardsley 1997).

The announcement of Dolly's birth generated a storm of media attention. It was not the cloning of a sheep, per se, that captured the popular imagination, but rather the potential use of this technology for cloning people. Within days of Wilmut's announcement, there were calls for ethical inquiries and new laws to ban the cloning of human beings (Pennisi and Williams 1997). While many scientists welcomed the opportunities that cloning might provide for gaining a new understanding of biological processes, religious leaders condemned what they saw as the latest sleight of hand though which science was attempting to tamper with the fundamentals of life. The Vatican denounced it. And the British government, swayed by an outpouring of negative public sentiment, announced an immediate end to Roslin's funding (Wallace 1997).

Central to the nature of the emerging debate are human embryonic stem cells, which were isolated in 1998. As many scientists have been quick to point out, cloning research falls into two distinct types—reproductive cloning, which aims to produce a fully-formed organism that is virtually identical to a currently existing (or previously existing) individual, and therapeutic cloning—a procedure that takes place at the microscopic level and is intended to study disease and cure human ailments (Dunn 2002). Although the two procedures start out the same (i.e., with a renucleated egg in a petri dish), researchers engaged in therapeutic cloning have no intention of creating a living organism. Instead, of implanting a reconstituted egg in a female host, their hope is to produce embryonic stem cells in a laboratory setting (Vogel 2002).

Stem cells are unique in that they have the potential to develop into any other type of body cell. Many scientists believe that these pluripotent cells hold great promise for understanding and treating many health-related conditions, including diabetes, Parkinson's disease, and several forms of cancer (Cibelli et al. 2002; Dunn 2002; Vogel 2001, 2002). A 2005 article in *National Geographic* describes the promise of this research in the following terms: "Few question the medical promise of embryonic stem cells. Consider the biggest United States killer of all: heart disease. Embryonic stem cells can be trained to grow into heart muscle cells that, even in a laboratory dish, clump together and pulse in spooky unison. And when those heart cells have been injected into mice and pigs with heart disease, they've filled in for injured or dead cells and sped recovery. Similar studies have suggested stem cells' potential for conditions such as diabetes and spinal cord injury" (Weiss 2005: 7). Some scientists believe that stem cells produced from cloned human embryos might be uniquely well suited for studying many diseases and devising novel therapies. In particular, use of tissue produced through therapeutic cloning helps to rule out problems of rejection commonly associated with transplant surgery. Several patient advocacy groups (including the Juvenile Diabetes Research Foundation, the Alliance for Aging Research, the American Liver Foundation, and the Kidney Cancer Foundation) and a host of biomedical organizations (including the Association of American Medical Colleges and the American Society for Cell Biology) have set forth impassioned pleas in support of this research (Hall 2002). This position has been countered by more conservative groups, who argue that such experiments create human life only to destroy it.

At the time of this writing, the United Nations committee in charge of international law has been unable to reach a consensus on whether to support a complete or partial ban on human cloning (Vogel 2002). Cloning for the purposes of therapeutic research is legal in several countries, including the United Kingdom, Singapore, China, South Korea, and the Netherlands. Other nations—Denmark, France, Austria, Canada, Spain, and the United States—have passed or are in the process of developing legislation that will prohibit the production of cloned human embryos for any purpose, including medical research (Vogel 2002: 1317; Weiss 2005: 7).

If the biomedical components of human cloning have elicited a mixed response, there has been unanimous agreement on the part of Western countries that human beings should not be cloned for reproductive purposes. To date there has been no country in either Europe or North America that has not explicitly condemned the practice of cloning to produce children. Although safety reasons are occasionally cited as a contributing factor, these are not seen to be the most pressing concerns.[3] Indeed, commentators generally agree that even if the technique came to be perfected through time, the practice of reproductive cloning would remain socially and morally unacceptable (Beardsley 1997; Brock 2002; Carlin and Biddle 1997; Cibelli et al. 2002; Dunn 2002; Hall 2002; Jaenisch and Wilmut 2001; Pennisi and Williams 1997; President's Council on Bioethics 2002; Segal 2002; H. Shapiro 1997; Vogel 2001, 2002; Wallace 1997).

In January 2002, President George W. Bush assembled an advisory council on bioethics to consider the potential benefits and challenges of human cloning. Six months later, the council presented Bush with a 194-page report outlining the results of their deliberations. Made up of academicians, lawyers, medical professionals, and a journalist, the council cited three main objections to reproductive cloning, above and beyond the health-related concerns; they claimed it would (a) threaten individuality and autonomy; (b) challenge the integrity of family forms; and (c) risk treating human children as objects of manufacture. The arguments set out in this report are interesting to consider. What they reveal (perhaps not surprisingly) is that the production of children through "normal" sexual means grounds more than Euro-American notions of "kinship." What is at stake, should reproductive cloning be allowed to proceed, are accepted notions of personhood, relatedness, and conventional understandings of human agency, including its limitations.

The greatest part of the council's report is taken up with a discussion of individuality and autonomy, and the presumed importance of these values to the formation of a human self. The concern that the council raises is that cloning would present "a unique and possibly disabling challenge to the formation of individual identity" (President's Council on Bioethics 2002: 102). This is because a cloned individual would inherit an identity "already lived in advance," and therefore would not fully be "a surprise to the world" (Kass 1997: 22). According to council members, this could cause profound psychological trauma. Throughout their report, the council is at great pains to point out that a "compromised sense of self" would not simply be attributable to genetic determinism. Identical genes, they note, do not make identical people (President's Council on Bioethics 2002: 103). What is at issue, however, are the expectations of "society." A cloned child would constantly be compared to the "original"—and this would impose undue constraints on his or her life course. As the problem is set out in the report:

> A cloned child is at risk of living out a life overshadowed in important ways by the life of the "original"—general appearance being only the most obvious. Indeed, one of the reasons some people are interested in cloning is that the technique promises to produce in each case a particular individual whose traits and characteristics are already known. And however much or little one's genotype actually shapes one's natural capacities, it could mean a great deal to an individual's experience of life and the expectations that those who cloned him or her might have. The cloned child may be constantly compared to "the original" and may consciously or unconsciously hold himself or herself up to the genetic twin that came before. If the two individuals turned out to lead similar lives, the cloned person's achievements may seem to be derivative. If, as is perhaps more likely, the cloned person departed from the life of his or her progenitor, this very fact could be a source of constant scrutiny, especially in circumstances in which parents produced their cloned child to become something in particular. Living up to parental hopes and expectations is frequently a burden for children: it could be a far greater burden for a cloned individual. The shadow of the cloned child's "original" might be hard for the child to escape, as would parental attitudes that sought in the child's very existence to replicate, imitate or replace the "original." (President's Council on Bioethics 2002: 103)

Richard Lewontin (2000) and others (Brock 2002; Segal 2002) have countered that identical twins are *more identical* to one another than a cloned organism is to the individual who donated the cell, in that twins share mitochondrial DNA in addition to other genetic material. Still, identical twins go on to develop a unique and autonomous sense of "self." The president's council, while acknowledging this point, goes on to argue that twins "begin life with a blank slate, equally ignorant of each other's identity," whereas a cloned child, by contrast, would start life with the knowledge of what his or her genetic predecessor has already become.

> Identical twins have as progenitors two biological parents and are born together, before either one has developed and shown what his or her potential—natural or otherwise—may be. Each is largely free of the burden of measuring up to or even knowing in advance the genetic traits of the other, because both begin life together and neither is yet known to the world.

But a clone is a genetic near copy of a person who is already living or has already lived. This might constrain the clone's sense of self in ways that differ in kind from the experience of identical twins. (President's Council on Bioethics 2002: 103–4)

If cloning is allowed to proceed, in the estimation of the council, genes, coupled with social expectations, will combine to undermine an individual's own sense of uniqueness and autonomy.[4]

A second set of concerns raised in the council's report is that cloning to produce children might prove to be "damaging" to family relations. They do not mean by this that cloned children would suffer from a lack of care and parental concern; rather, the child's place in the scheme of family relations would become "confused" and "uncertain." In the council's opinion, the "usually clear" designation of kinship relations that emanate naturally from the genealogical grid would fall into hopeless disarray. The role of individual family members would become chaotic and blurred. What is ultimately at issue, here, is the seemingly "unnatural" condition of having to relate to one person (the clone) as both "self" and "other": "By confounding and transgressing the natural boundaries between generations, cloning would strain the social ties between them. Fathers could become 'twin brothers' to their 'sons,' mothers could give birth to their genetic 'twins,' and grandparents would also be the 'genetic parents' of their grandchildren. . . . Every other family relation will be similarly confounded" (President's Council on Bioethics 2002: xxix). In a separate publication, Dr. Leon Kass (1997), who headed Bush's council, likened self-cloning to incest, in that it results in a situation whereby an individual becomes a parent to his or her "sibling" (23). What should be kept separate, both "genetically" and "socially," comes to be conjoined in an unnatural and potentially dangerous union.

Finally, the council highlights what it sees as one additional concern: "the transformation of human procreation into human manufacture" (President's Council on Bioethics 2002: 104). Cloned children, not unlike other objects of intentional human design, would be brought into being with a predetermined pattern in mind. In "natural" procreation (and most forms of assisted conception), a child is created through the random combination of parental gametes. Traditional conception entails a "genetic lottery" of sorts, in that the end result is a matter of happenstance (Brock 2002: 314). Cloning to produce children is different. With cloning, parents set out to produce a child with a specific genotype—that of the somatic cell donor.

Cloning thus allows one to determine in advance the precise genetic composition of one's offspring. The implications of this new technology, according to the council, are both momentous and frightening:

> The principle that would be established by human cloning is both extensive and completely novel: parents with the help of science and technology, may determine in advance the genetic endowment of their children. To this point, parents have the right and the power to decide *whether* to have a child. With cloning, parents acquire the power, and presumably the right, to decide *what kind* of child to have. Cloning would thus extend the power of one generation over the next—and the power of parents over their offspring—in ways that open the door, unintentionally or not, to a future project of genetic manipulation and genetic control. (President's Council on Bioethics 2002: 105)

The council's *Human Cloning and Human Dignity: An Ethical Inquiry* is intriguing to consider for several reasons. For our present purposes, one of the most fascinating features to note about the document is the extent to which it valorizes and fetishizes a particular view of self and social life that is attendant upon a biological paradigm.

Briefly, I wish to draw attention to three important assumptions that underlie the logic of the council's report. First, although the degree to which genes influence identity is seen as debatable, what the council takes as axiomatic is that selves are expected to be unique. Each human life is understood to have a quality about it that is not replicable. Whether this is a product of genes or the environment—of "nature" or "nurture"—is largely irrelevant in their account: what is taken for granted and beyond needing to be questioned is the "essential" nature of our individuality (cf. Battaglia 1995, 2001). Second, while the council attempts to avoid the pitfalls of genetic determinism in their discussion of personal identity, biology is seen to play a determining role in the definition of social relations. The claim that a cloned child's grandparents are technically also his parents makes sense only to the extent that genes are understood to define parentage. Third, the production of life and sociality is seen to be characterized by randomness and unpredictability—neither is subject (or, at least, should be subject) to conscious human control. Replication is understood to be both undesirable and unnatural: it is the constant production of "newness" that comes to be positively valued in this model.[5]

In the remainder of this chapter, these images will be placed up against

Kamea conceptions of the world. Unlike Europe and North America, where the production of unending novelty and diversity is something to be celebrated, throughout much of Papua New Guinea the aim of social action is precisely to replicate certain identities and relationships. This point was understood early on by Maurice Leenhardt, the anthropologist and missionary whose classic work on New Caledonian culture first appeared in 1947. In a discussion of Canaque naming practices, Leenhardt demonstrates that New Caledonians consciously strive to re-create particular identities and relations across the generations:

> The ancestral name, periodically restored over the generations, actualizes the former personage by investing a new person in the society with his august personality. . . . Homonyms offer the perfect example of this investiture. Caledonians consider a man to be a namesake only when he bears all the names the elder had, with—legally, but not now in fact—all the participations which these names may imply. The totality of names is necessary for reconstituting an identity. By virtue of this identity, homonyms treat one another as brothers, with equal legal rights: each is a replica of the same personality. (Leenhardt 1979: 156)

Below, I explore this issue in greater detail by turning to an examination of death practices among Kamea. My aim in this discussion is twofold. First, I will demonstrate that Kamea mortuary practices are explicitly geared toward reproducing certain identities and social relationships through time. A second goal of this chapter is to assess the adequacy of existing anthropological interpretations of death. More specifically, I shall show that orthodox approaches to mortuary rites take a genealogical framework as their analytical baseline. Thus, Western notions of biology and all that it entails have heavily influenced not only how we see the generation of life—they have also structured how we view what it means to die (cf. M. Strathern 1992a). In the pages that follow, I highlight the extent to which a biological model has impeded our ability to grasp what the termination of life might mean in a non-Western context.

DECONCEIVING ETHNOGRAPHIC ASSUMPTIONS

Anthropologists have long looked to mortuary rites as an analytical springboard into those cultural values that underwrite the construction of social life. As Huntington and Metcalf (1979) argued more than a quarter of a century ago, death throws into relief important ideas through which people live their lives and evaluate their experiences: "Life becomes transparent against the background of death" (2). Given the enormous range of variability that characterizes social life the world over, it is somewhat surprising that one model of death has managed to dominate ethnographic accounts. From nearly the beginning of our discipline to the present time, anthropologists have framed their interpretations of mortuary events in terms of explanations that relate funeral practices to the reproduction of an existing social order (Bloch and Parry 1982; Frazer 1911; Hertz 1960; Huntington and Metcalf 1979). Death, we are told, is tied to the renewal of society, thereby linking images of social reproduction with those of decomposition and decay. Bloch and Parry (1982) summarize the received wisdom of this position when they write: "The renewal which occurs at death is not only a denial of individual extinction, but also a reassertion of society and a renewal of life and creative power" (4).

Anthropological treatments of Melanesian mortuary symbolism have fit rather well with orthodox interpretations of death. Indeed, in their approach to death, scholars of Melanesia have exhibited a conservatism that runs counter to their more celebrated "deconstructive" bent. I do not mean by this that studies of death in Papua New Guinea have simply aped what has been written on the subject elsewhere. Melanesianists have added to the discussion a concern with the manner in which exchange accomplishes at a practical level the stated aims of mortuary rites—namely, "social renewal." In a cultural milieu that is well known for its exaggerated emphasis on exchange (Feil 1984; Meggitt 1972; Sillitoe 1979; A. Strathern 1971b; Wagner 1967), the efficacy of mortuary rites is often claimed to be tied to indigenous interpretations of reciprocity. By carrying out what frequently amounts to extremely complicated exchanges at death, Melanesians, anthropologists have argued, legitimate, and in some cases create, the categories upon which social life is based.

Robert Foster's (1995) work with Tanga Islanders serves as an eloquent illustration of the application of this argument. In his analysis, Foster argues that the sequence of mortuary feasts that Tangans enact in the wake of death help to elicit collective identities in the form of matrilineages. By manipulating what is exchanged at death (durable vs. nondurable forms of wealth) and the roles that are played by participants in the exchange (i.e., whether one serves as a feast giver or a feast receiver), Tangans transform death into a realization of matrilineal continuity:

Specifically, I argue that by hosting a sequence of mortuary feasts, a group of enatically related men and women realizes for itself the identity of a collective individual impervious to the exigencies of consumption, namely death and decay. Ceremonial transactions in which the host's exchange cooked pigs and food for their guest's shell valuables catalyze this realization. By eliciting from their guest's gifts of shell valuables, iconic signs of permanence and durability, feast givers effect a kind of asexual regeneration. That is, feast givers directly appropriate the qualities of these objects as definitive aspects of their collective individuality. Ceremonial exchange emerges as the "real" source of matrilineal continuity. (Foster 1990: 432)

Foster's analysis, like others in this theoretical vein, is based on a specific view of what constitutes the social person in Melanesia: a view that is widely supported by the ethnographic literature and has its roots in what has been reported of indigenous ideas concerning the heterosexual nature of reproduction. As I noted in chapter 2, for many Papua New Guineans, maternal and paternal contributions at conception are said to be given equal weight and play an equally important role in defining an individual's social field. Indeed, social life frequently plays itself out in terms of managing the competing claims of bilateral kin and compensating one side—either maternal relatives or paternal relatives—for their role in helping to create the body of a child. Because various components of the nonhuman environment serve as an external and partible adjunct to human bodily fluids (A. Strathern 1973; Wagner 1977: 631; J. Weiner 1982: 9), gifts of substance can be bestowed upon an individual after birth, thereby magnifying and expanding relations of indebtedness. Death, then, becomes a time when "accounts" are finally settled: it exists as a moment in time when this "composite" (M. Strathern 1988) form is "decomposed"—both "socially" and "physically"—and its constituent parts are sent back to the places from which they came.

> Feast-givers symbolically return to feast-receivers the bloods of women that had come inside the clans of feast-givers from outside with the procreation of the deceased's parents. By sending these bloods back to where they originated, the deceased and through him the feast-givers collectively are deconceived from these specific cognatic kin. (Mosko 1983: 31)

> A death must be understood as the moment when huge amounts of restoring occur: when all retrievable elements (including the deceased's body and bones) are reordered and redefined by *dala* owners. In this respect, death triggers the return of the deceased's body and property to members of its

own *dala* followed by the concern to make the loss of the deceased (and the loss of the social relations) into a positive resource for the regeneration of equivalent social relations. (A. Weiner 1980: 81)

Gawan mortuary exchanges are concerned with factoring out the marital, paternal and maternal components which have been amalgamated to form the deceased's holistic being, and with returning this being to a partial, detotalized state—its unamalgamated matrilineal source. (Munn 1986: 164)

Two ideas underpin this system of representations. The first is that human sexual reproduction forms, at least in the first instance, the underlying basis of kin connections. Second, implicit in these descriptions is the idea that the so-called "physical body" and the "social body" (i.e., "society") are analogues of one another (Douglas 1970: chap. 5). By manipulating the body of the corpse, the social group of the deceased is said to be given an opportunity to reclaim those resources that have been lost through previous acts of procreation, and to reconstitute itself as a reproductively viable and differentiated social entity. As Mosko (1983) notes, using wealth as a substitute for bodily substances, the constituent parties to the exchange "redefine or close their mutual [social] boundaries" (27).

Kamea represent a particularly compelling ethnographic case to consider—both in terms of general anthropological theorizing on death, and with respect to how it gets worked out in its specific Melanesian incarnation. As we have seen in previous chapters, Kamea do not imagine bodies as rigidly bounded entities; instead, they pass back and forth between singular and composite states. We have also seen that bodily substance plays only a negligible role in defining social relationships and that lineal continuity bears little resemblance to anthropological notions of "descent." Different views of life and ultimately of death are entailed by these views.

Given this, an interesting problem presents itself. How is the loss of the social person through death experienced when the process of reclamation is not an issue: that is, when there is no ready-made "place" for things to get sent back to, and where the point is to *create* social ties rather than *reproduce* them? In what follows, we shall see that Kamea funeral practices are not centered on disarticulating bodies and the relationships of which they are composed to yield an exclusive social identity, but rather, on transforming one type of social relationship into another. More specifically, they capture the ongoing dialectic through which relationships based on siblingship and affinity eventuate from one another. As we shall see, cross-cousins are central in this transformational process.

I first became aware of the importance of Kamea death practices almost immediately upon arriving in the mountains of Gulf Province. I had been at the government station at Kaintiba for about a week and was still searching for an appropriate field site upon which to base my research, when Dabenrette—the local woman who had been acting as my guide and interpreter at the time—arrived one day to tell me that she would not be able to assist me that day: her "nose" (*hiyma*) had died, and as a consequence she had an endless array of obligations that required her immediate attention. With a bit of prodding, I soon learned that Kamea liken one's cross-cousin to one's nose, and that the relationship between cousins is intense and encapsulates a broad range of meanings. My interest was thus piqued, and for the next two years I tried to learn all I could about the association between cross-cousins, noses, and death.

When an individual—either male or female—dies among Kamea, news is circulated as quickly as possible to all of the friends and relatives of the deceased. The body is then placed in the home of an agnatic kinsman (typically a brother, son, or father), where it remains on display for several days while a period of mourning is undertaken. The length of time that the corpse is available for viewing is highly variable. Christian missionaries have been strenuous in their campaign to have the body buried as soon as possible after death, fearing that prolonged periods of mourning promote the erroneous idea that the spirit (*hikwapa*) of the dead remained with living, rather than residing in the hands of God. Kamea, for their part, prefer to hold longer viewing times as a means of ensuring that all who wish to come and view the body have the opportunity to do so.

As the body lies in state, it is tended by various kinsfolk. The immediate family of the deceased is the most notable presence. Siblings, children, spouses, and parents crowd into the house of mourning, where they lament and wail throughout the night. Emotional outbursts are common during this time. People may throw themselves upon the body of the deceased or lash out at others, particularly if sorcery is suspected as a cause of death (cf. Wagner 1972: 145).[6]

While the body rests in the mourning house, friends and relatives of the deceased create memorial songs (*am'a pit'ya apa:* "man-dead-song") that celebrate the now finished life of the deceased. These frequently recall significant events in the dead person's life or remember other aspects of his or her existence, such as their appearance or the vigor with which they worked a garden while alive. The most common type of song that is performed on these occasions draws heavily upon images of the landscape (cf. Schieffelin 1976: 177–88). As we saw in chapter 1, Kamea constructions of identity are intimately tied to their surrounding rain forest environment. Mourning songs often proceed as a journey, in which the singer calls out a number of places that the deceased frequented during the course of his or her life. These moving expressions of grief chronicle the life of the deceased by using the landscape as an objectification of social and personal experiences.

Those who experience the loss most acutely adopt visible signs of their aggrieved condition. Using a mixture of soot and water, they blacken their bodies to signify their "emotional heaviness." As one man explained: "When someone has died, we take the ashes from the fire and mix it with water. Then we put it [the mixture] on our skin. We are sorry for the person who died, so we take earth with ashes and put it on our bodies. Afterwards, we walk about all black."

Full mourning attire consists not only of a blackened body, but also a mourning ornament known as *he'aka,* made by rolling ficus tree bark into strings that are draped about the shoulders and neck. Shorter strings are attached to the main holder lines and hang down to form a heavily fringed bib. Personal effects of the deceased, a cutting from a favored laplap, the skin of a betel nut that he or she chewed, are weaved into the individual strands. Like mourning songs, these ornaments objectify specific memories of the deceased, which gradually dissolve, a process Kamea say is coterminous with the rotting of the *he'aka* themselves, which are hung on trees after the final mortuary rites have been held (this point is taken up in further detail below).

This kind of remembering is also emphasized in the selective food taboos that mourners may take on in memory of the deceased (cf. Munn 1986: 170; Schieffelin 1976: 64). Fruit of the pandanus tree (marita) and food that has been cooked in an earth oven (mumu) are avoided by those who wish to emphasize a particularly close relationship to the deceased. While this often includes siblings, spouses, and children of the deceased, the decision to adopt these interdictions is voluntarily undertaken. The same applies to the growing of one's hair in the weeks and months immediately following death, which serves as another unambiguous sign of mourning. It is only when the final mortuary rites are held approximately two years later that the process of forgetting the dead officially begins.

Preburial grieving continues over a period of several days. During this time, people pour into and out of the house of mourning. The corpse is

Figure 9. A woman wears a *he'aka* to commemorate the recent death of a kinsperson. Photo by Sandra Bamford.

never left alone during this time. Witches (*pannga*) and a wide variety of bush spirits (*hiey'ya*) are said to find the taste of the newly dead particularly "sweet." The presence of mourners helps to dissuade these beings from engaging in their cannabalistic feasts. The corpse is typically buried two to three days after death in one of the small cemeteries that are situated on the outskirts of the village. Anyone who wishes to attend the graveside ceremony is free to do so. The dead is placed in a wooden coffin and carried by friends and relatives to the place where he or she is to be interred. A brief Christian ceremony will be performed while the casket is being lowered into the ground and the hole filled with dirt. Personal effects of the deceased are often left to adorn the gravesite. When the service has been completed, those in attendance disperse.

Several years later a feast in honor of the recently deceased (*am'a pit'ya itya iymakio:* "man-dead-food-make") will be held. This constitutes the final phase of Kamea mortuary activities. It is the deceased's bilateral cross-cousins who will act as the official recipients of the feast. Even more than a parent, child, sibling, or spouse, bilateral cross-cousins are said to be particularly inconsolable when it comes to the death of their kinsman. They typically vent their anger toward the immediate family of the deceased (a spouse, a parent, or in some cases a sibling), who are accused of negligence for allowing the dead person to die. Overcome by the intensity of their emotion, they may express their ire by breaking arrows, ruining gardens, cutting down fences, and destroying houses. A cousin's wrath will eventually prompt them to approach the immediate family of the deceased and demand to be compensated for their overwhelming loss. The food given at death feasts is said to "cool" (*ipakati*) the wrath of cousins and to restore a sense of emotional equilibrium (see below).

THE PRECOLONIAL CONTEXT

What I have described thus far are the mortuary rites as I witnessed them at Titamnga during the early 1990s. Prior to the arrival of government and missionary agents during the 1960s, the situation was somewhat different, and these differences help to illuminate the meaningfulness of Kamea death rites.

In the past, when a man or woman died, a huge platform would be built, upon which would be placed the body of the deceased. Large posts made from banana palm trunks (Simpson 1953: 164) would be set vertically into the ground and a series of smaller planks would be laid horizontally over the top. The deceased was then placed in a seated position on top of the scaffold with arms extended at the side and feet fastened to the platform with a rope. Several small smoking fires would be lit nearby, tended by onlookers on a continual basis. The heat from the fire would gradually dry the corpse and prevent the skin and muscles from fully decomposing. When the smoking procedure was complete, the mummified remains of the deceased would be placed in nearby rock shelters and caves.

The drying process generally took between one to three months to complete, depending on the size and weight of the body.[7] During this time, cross-cousins of the deceased were expected to sit underneath the corpse to catch the fluids that dripped from the body (cf. Mbaginta 1976). The heat from the fire would cause the skin of the deceased to blister, and it was the duty of cousins to rub their dead kinsman with bark cloth in order to help facilitate the drying process. As the secretions fell from the body, those sitting underneath would smear themselves with the fluids of decomposition until they were thoroughly saturated.[8] Throughout this time, cousins would not handle food, except with a spearlike implement called an *ikatanga,* in order to avoid directly ingesting the fluids of the dead.[9] When the corpse was completely dehydrated, it weighed about thirty pounds (Simpson 1953: 165) and could be brought to its final resting place to be interred within limestone cliffs. My Kamea consultants said that it was in return for their service as undertakers that cousins received things at the mortuary distributions that took place several years later (see below).

Although the practice of smoking corpses had been abolished long before I arrived in the district, it is possible to find a general account of it in the ethnographic writings of Beatrice Blackwood (1978). As noted earlier, Blackwood conducted research with a Kapau-speaking group in the 1930s, roughly four days walk due east of where Titamnga is situated. Her description is one of the few firsthand accounts of traditional death rites in the Kamea region.[10]

When I was at Andarora I accompanied some mourners to a smoking. On 23 December a crowd of men came up, and Pukpuk said they were going to cry over the man who died the other day at Gunugunauka hamlet. I asked if I could go, and there seemed to be no objection; Sitoun said he would come too, but changed his mind and said he was afraid the tamboran (spirit) of the dead man would eat him, so Nimo, his younger brother who talks some Pidgin, was deputed to look after me.

We set off to the house on the north hill with three men and about half-a-dozen women, all the men with bows and arrows, and some with adzes, while some of the women carried babies, and the party was also accompanied by some small boys. Sounds of wailing came down the hill before we reached the house; as we approached, our party started wailing, and just before going in the house each man took one or two arrows from his bunch, broke them and threw them on the ground. They were afterwards picked up by a small boy and stuck into three banana stalks leaning against the roof. The other arrows, and the bows, were leant against the wall outside. The hut was a round one, with a pent-house roof, the door only about 1 ½ x 2 ½ feet. We went in one by one. I waited till the last except for Nimo, my guide and interpreter, who followed me in.

The hut was packed with people; I had just room enough to half sit by the door, since the roof was not high enough for me to stand upright, and I had to lean against the wall. For some time I could see little in the darkness and smoke, but finally made out a corpse at the back of the hut facing the door. No one took any notice of my entrance, except the woman next to the door, who moved an inch or so and motioned to me to sit down.

The women sat on the left of the corpse, the men on the right, except for one woman beside me with a baby. The women wore cloaks, but the men didn't. The body was that of a boy in his late teens, but so bloated that it was unrecognizable as such until I was told. It was placed in a kind of swing chair made of cane, suspended from the roof, the bottom about a foot from the floor. Beneath was a layer of ashes, and on these some grass or leaves. In front between the body and door was a fire, the smoke of which filled the hut and seemed to go in every direction except on the body. The body was in a sitting posture, its arms extended forward, slightly bent at the elbows with the hands hanging down, the head upright, with mouth and eyes open, tongue between the teeth. Thongs were fastened round the chest under the arm-pit and just above the groin, very tight so that the belly swelled out. There were also thongs around the knees and ankles, fixing the feet to a square of cane below. The smell was overpowering.

People were packed so close that I could hardly move my hand to my breeches pocket to get a much needed handkerchief. It must certainly have been the first time a white person had wept at a Kukukuku funeral, though it was smoke and not grief that filled my eyes with tears. The wailing was incessant, sometimes rising to a higher pitch; the words were indistinguishable, and seemed to be simply *awe awe*.[11]

The boy was not married, and his mother lived a long way off. Two of his sisters sat close to the body on the left and rocked the swing gently. A man squatting behind came forward and put more wood on the fire. Pieces of firewood were stacked up against the house on both sides of the door—boys handed them in from time to time.

I stood it as long as I could, and when I hoped it was long enough to satisfy the claims of etiquette I took advantage of the fact that there was not enough room for a man who looked in at the door, to come out and sit on the ground outside with a crowd of men, women, and children. They were chewing betel, discussing pigs, laughing, and not particularly sorrowful, and interested in the spikes on my boots, my writing these notes, etc. Presently Nimo wanted to go, and I didn't try to stop him. The others would stay until late afternoon. This would go on for about a month till all of the "water" had gone from the body and they could put him among the pandanus trees. (Blackwood 1978: 134–35)[12]

Blackwood's account of the smoking process is similar to that described to me by the men and women of Titamnga, although there are some important differences. In the western Kamea region, of which Titamnga forms a part, cross-cousins were expected to sit underneath the body of their kinsman and to anoint themselves with the fluids that fell from the body. Several of my older consultants possessed distinct memories of having performed this service in the past, before the Australian colonial government outlawed the practice.

As Mimica (1991) has pointed out about the neighboring Ikwaye peoples, the subject of smoking the dead is a highly sensitive one among local people (92). Missionaries have had considerable success in convincing local peoples that their traditional way of handling the dead is to be despised. Indeed, it was only after I had been at Titamnga for about a year that people began to discuss the subject with me. Yet, although the practice of smoking corpses no longer exists, people continue to demonstrate a potent concern with the social obligations that death engenders. As I argue below, the meaningfulness of funeral practices among Kamea extends far beyond dealing with the physical remains and grieving the loss of a kinsman.

Kamea say that they smoked their dead in the past so that they would stap gut—that is, so that the body of the deceased would remain intact and not decay as inevitably happens with modern-day burials. "We smoked them so that they would not rot quickly. Once dry, we would place them in stone [caves]. The man would be like 'new' (nupela). He would stay like new." When the smoking process was complete, the corpse took on a shrunken and leathery appearance. In this form, it would be carried to a limestone outcrop where it was placed alongside the bodies of already dead paternal kinsmen. Ensconced within its stone tomb, the mummified figure of the dead remained for decades, an objectified and enduring testament to the care that a cousin took for one's well-being, not only in life but in death as well.

DURABLE BODIES, FLUID RELATIONS

At first glance, Kamea mortuary practices appear as though they might fit rather well within the parameters of our current analytic paradigm. Indeed, when I first sat down to make sense of what the men and women at Titamnga had told me, indigenous practices appeared to be grist for the Durkheimian interpretative mill. Kamea death rites, I managed to convince myself initially, were about disassembling a composite being and sending its constituent elements back to the places from which they came. By sitting underneath the body of their dead kinsman while it decomposed, cousins effectively separated the perishable and imperishable aspects of existence—elements that are commonly associated in the literature of Papua New Guinea with female and male—so that the more enduring male components could be placed in caves as a symbol of lineal continuity. By emphasizing the removal of (female) substances, Kamea death rites produced a singular social form from a prior act of bisexual reproduction.

At the same time that I was formulating this idea, I was growing increasingly uneasy about it. How could mortuary rites be focused on the removal of "female" substance when women were not seen as sharing bodily substance in common with their children to begin with? In a world where the parent-child tie is not conceptualized in terms of the inheritance of biogenetic substance, removing the "female" contribution to conception made little sense. Furthermore, unlike those systems where a conceptual distinction is made between flesh and bone, and where the two components of the person are subject to markedly different treatments at death (see, for example, Bloch and Parry 1982; Hertz 1960; Huntington and Metcalf 1979), Kamea mortuary practices had as one of their primary aims *preserving* the flesh of the corpse rather than allowing it to disarticulate. Here, every effort was made to arrest the process of decay, rather than allowing it to assume what Euro-Americans see to be its "natural" course. Kamea death rites, I began to realize, were about something else. To grasp their significance, it was necessary to adopt a different perspective.

Kamea ideas about how social life unfolds through time differ significantly from those upon which the dominant model is based. If orthodox anthropological theory uses the "facts of biology" to speak about the connection between what are seen to be innately differentiated bodies, Kamea presuppose a state of relatedness, and work to define different types of relationships within this underlying social field (cf. M. Strathern 1988; Wagner 1967, 1975, 1977). As we have seen in previous chapters, cross-cousins are central to this process of differentiation. Kamea emphatically state that first-degree cross-cousins should not marry; however, the children of bilateral cross-cousins are enjoined to do so. This means that it is cousins—themselves one step removed from the similitude of "one-blood" relatedness—who sit down to arrange a match between their respective offspring. Cousins, in short, occupy the space between two containment cycles. Through their work in negotiating marriages, an affinal bond is brought into being, and at the same time the similarity of "one-bloodedness" is replaced with a cross-sex sibling tie. Cross-cousins are situated at the beginning and end of a process wherein a host of other types of identities and relationships are conceived.

The traditional practice of smoking the dead serves as a metaphor of Kamea understandings of what connects and disconnects bodies in a world where substance is not used as a means of tracking social relationships through time. It also speaks to the fact that as a pivot between different modes of relating, cousins are both similar and dissimilar to other categories of kin: they furnish the "raw material," so to speak, from which other types of social relationships are engendered as different ways of being in the world. The treatment of the body at death in the Kamea world reveals precisely this same tension between unity and differentiation. Instead of disarticulating the body into a series of constituent parts and subparts, they present an image of simultaneous unity and separation. On the one hand, every effort is made to preserve the material integrity of the corpse so that it will <u>stap gut</u> (literally, "stop good"), as Kamea put it, rather than fall completely out of existence. This can only be achieved, however, by removing the cadaverous fluids from the corpse, an act that is simultaneously disarticulating. Preservation and decomposition undergird life and death. Cousins are central to how both of these processes are realized.

Sitting underneath the body of the corpse while it is being smoked is an appropriate activity for cousins who serve as the fulcrum between relationships of siblingship and affinity. Just as the parents of cross-cousins were once contained within the same prenatal environment, so too do cross-cousins come to share a similar environment within the context of Kamea death rites. Sitting beneath the corpse while it decomposes places cousins in a position analogous to that of "one-blood" siblings. They come to share with one another a similar environment: their joint containment by the fluids of decomposition. What is important here are not the cadaverous fluids in themselves, but rather, that those who are doing the sitting share with one another a similar context: they are all encompassed or contained by the same juices that are dripping from the corpse. In the same way, "one-blood" siblings are not united by sharing substances in common; rather, what they share is a similar context by virtue of the fact that they were all contained within the same woman's womb. It is context as opposed to item (i.e., substance), relationship as opposed to essentialized identities, that furnish the ground upon which Kamea sociality is based.

But, if cousins are "like" siblings by virtue of their positioning vis-à-vis the corpse, they are also different, just as their intermediate position in the kinship system emphasizes. At conception, male and female procreative substances mix within the womb of the mother. It is the joint containment of these fluids that leads to the creation of a child and that subsequently relates children who have been born of the same womb as "one-blood" siblings. At death there is an inversion of the containing/contained relationship. Here, it is not bodies that come to act as the container of fluids, but rather fluids that do the containing of bodies. I noted earlier that cousins exercised extreme care during this time so as to not inadvertently ingest the fluids of decomposition. All food was consumed with a special several-pronged spear so as not to pollute those items that were to be taken within the self. Through their participation in death rites, cousins turned everything inside out so that what was once contained comes to do the containing instead. In the generation that immediately follows, the cycle begins anew when the children of first-degree cousins marry and produce a child through the recontainment of their differentiated sexual fluids.

It is significant in terms of the overall imagery of the rites being examined that cadaverous fluids are never spoken about in gendered terms. If conception is achieved through the coming together of male and female procreative fluids, it is the unsexed bodily products of the deceased that furnished a containing environment at death. What is seen as the unification of duality at conception is decontained at death in a composite form. In its capacity to symbolize the mutual embeddedness of "kinship" and "gender" relations, what takes place at death exists as an analogue of how life is created.

For Kamea, the ongoing oscillation between unity and differentiation reflects more than the nature of existing social processes—it exists as a fundamental principle through which the world came into being in its present form. It will be recalled that the myth of origins recounted in chapter 1 has multiplicity emerging from a state of homogeneity. Akeanga's wives set fire to their house with their husband still inside and later deposited his charred remains in a pool of water. There, his body underwent a series of remarkable transformations, changing first to tadpoles, then into partially metamorphosed human beings, then into male initiates, and finally into mature human beings. From the singular source of Akeanga's body springs forth the entirety of humankind. It is the sisters' act in assigning names to men as they spill from the tree that breaks this primordial unity into distinct social categories. The myth also introduces a second important theme: it shows how the world is founded upon a series of transformations (bones become tadpoles, which become human beings, etc.), rather than a stable set of enduring identities. In Kamea thought, it is transformation and change, rather than fixity, that is the assumed baseline for existence.

Orthodox treatments of death rest on the supposition that the "social" and the "physical" body exist as metaphors of one another. Because substance possesses the capacity to move beyond the margins of the body, it has traditionally been seen by anthropologists as an apt mediator of "individual" versus "social" existence (Douglas 1966: 114–28). Consequently, how bodily products are handled at death has been seen to become an important means by which the constituent units of society are buttressed and given a transcendental value. The images manifest in Kamea death rites, by contrast, give the opposite message. Here, those fluids that transverse the boundaries of the deceased are rubbed on the skin of another—his or her bilateral cross-cousin. These fluids become a medium of *identification* with the other, revealing not differentiation, but the existence of a potentially undifferentiated state. While the category of cross-cousin denotes a very specific relationship for Kamea, the uniqueness of this tie derives from the fact that it encapsulates multiple types of identity within it: siblingship and affinity; male and female; and the living and the dead. Death becomes a moment at which the fundamental units of Kamea social life fold in on themselves to reveal the existence of a world that is not based on innate distinctions, a world where cousins act simultaneously as totalizing and detotalizing figures.[13]

Kamea cross-cousins are the point of articulation between two types of cross-sex relations: siblingship and affinity. Although first-degree cross-cousins cannot marry, their children are expected to do so, and in the process reestablish the conditions of lateral relatedness. Significantly, this movement toward "recontainment" takes place not only at the level of those offspring that the newlywed couple produces, but also at the level of the original sibling unit. My consultants at Titamnga confirmed Beatrice Blackwood's important observation that women, upon death, are repatriated to their natal place (Blackwood 1978: 135). Instead of being smoked (or, nowadays, buried) by her husband's people, a woman's corpse will be brought back to its place of origin for disposal. At death, then, siblings— the product of a like containment—are brought back together again to be recontained within the same stone tomb. The dried corpses of brothers and sisters are encased together for eternity. Having been differentiated during the course of the life cycle, "one-blood" similitude is finally brought back together at death.[14] Cousins, through their work as undertakers, recombine the sibling unit just as their own offspring in life are brought together through marriage and affinal relations. A perceived "sameness" between alternate generations (see below) is expressed through its like containment. It is tempting to argue that the recombination of siblings at death is necessary for the production of new social relations: it is only after the final mortuary rites have been held that the widow of the deceased is free to marry again. The recontainment of siblings marks a return to a state of unity.

Kamea social life is founded upon the dual processes of containment and decontainment. The exchange of services between cousins at death makes possible the replacement of alternate generation relations. Grandparents and grandchildren employ reciprocal terms of address; they are also brought into a containing relationship through the activities of cross-cousins. The manner in which the body is treated at death reflects the different sides and potentialities of social life. Cousins remove from the corpse the fluids of decomposition, but at the same time they act to preserve the integrity of the "physical remains"—they serve as an image of closure and potentiality, ephemerality and durability. In the next section, I will show that the manner in which cousins are compensated for this service is entirely appropriate for the role they play in social life.

FEASTING AND THE METAMORPHOSIS OF SOCIALITY

Alongside initiation, mortuary feasts (*am'a pit'ya itya iyamakio:* "man-dead-food-make") are the most public and elaborate ritual events to be carried out by Kamea. These feasts are held one to two years after death and have as their aim to "finish" (*ipakati*) (or more appropriately, to inaugurate a process of "finishing") all thoughts of the deceased. The actual timing and scale of these events depends on a number of factors, including the availability of resources: the supplies of sweet potatoes, taro, pig, and, these days, rice and tinned fish; the availability of key personnel and their access to both food and money; and the history of relations between the deceased and the living—particularly those who will serve as feast givers and recipients (cf. Battaglia 1992). In some cases, the circumstances under which the deceased died also figures in the timing of the event. In one feast that took place at Titamnga in 1990, foodstuffs were presented to cross-cousins of the deceased (the official recipients of these distributions) only six weeks after the death had occurred because feelings ran high in the wake of a particularly virulent sorcery dispute.

In the majority of cases, the closest living male kinsman of the deceased will act as the principal sponsor of the feast. Say, for example, that a middle-aged man with a wife and children dies. If his son is already grown and has a family of his own, he will characteristically act as the official host of the distribution. However, if the son is too young to fulfill the obligation, a brother or father will step in to assume the duty instead. When a woman dies, her death feasts are generally performed by her husband or by a cross-sex sibling. Although Kamea never phrased it to me in precisely these terms, the expectation seems to be that the duty generally falls upon that person who was most closely involved with the care of the deceased during life. Mortuary distributions are not held in the event of the death of a young child, a point to which I shall return later in this chapter.

As noted earlier, it is bilateral cross-cousins who act as the chief beneficiaries of mortuary feasts. These payments are said to cool and contain the anger that the loss of a cross-cousin inevitably produces. Several people also suggested that, at least in the past, cousins could demand things at death as recompense for all of the work that they did in preparing the body of the deceased. Yet, it is significant that cousins have retained their role as principal feast takers despite the coming of Christian burial, a telling sign that their importance in death rites extends far beyond their traditional role as undertakers.

As Kamea conceptualize the process, "finishing" the dead does more than restore a sense of emotional equilibrium. The exchanges also occasion an opportunity to actively remember those who have died. I pointed out earlier that shortly after a death has occurred, the friends and relatives of the deceased will manufacture *he'aka*—heavily fringed mourning ornaments—which are worn about the shoulders and neck and later stored in the house of the bereaved when he or she tires of wearing them (see figure 9). When the final mortuary distributions are held several years later, these ornaments will be brought outside and hung in the branches of trees, where they will be allowed to decompose and eventually become part of the ground. The disintegration of these ornaments through normal processes of decay is said to parallel the process of forgetting the dead. As Munn has remarked of the symbolism of Gawan mortuary rites, they "involve the creation of a temporary memorialization so that paradoxically forgetting can be generated" (Munn 1986: 66).[15]

As the appointed day approaches, the "owner" (or papa) of the event will begin to check and recheck what resources are available to him. He will attempt to locate from within his own holdings a pig of the appropriate size that can be given as the pièce de résistance of the mortuary exchange. The most common reason for not fulfilling one's mortuary obligations in a

Figure 10. Residents of Titamnga gather for a village feast. Although most people spend the bulk of their time at garden houses in the bush, feasts provide an occasion upon which large numbers of persons come together. Photo by Sandra Bamford.

timely fashion has to do with the lack of a suitable beast. Should a man find his own herds wanting, he will scour the holdings of close kinsmen, often walking through miles of bush, to find a pig that is suitable for the event. Wives and sisters will also redouble their gardening efforts during this time in order to ensure that enough produce is available to feed all of those in attendance. As the feast organizer prepares, he will call in whatever monetary debts are owed to him so that cash is available to purchase rice, tinned fish, and other store-bought commodities. Those who wish to contribute to the "strength" (*yanga*) of the feast will leave donations of food or money with the organizer. The borrowing and lending of these social necessities is remembered and reciprocated in kind at a later date.

On the day of the feast, the chief sponsor and his helpers carry all of the assembled food items to a previously designated location. Ordinarily, the feast is held at the home of the eldest male cross-cousin of the deceased. However, should this person live some distance away, a younger cousin will act as the principal recipient instead: it is the existence of an active relationship rather than age alone that determines who will act as the recipient at these distributions. The primary feast taker will accept the food that has been offered, not as an individuated agent, but on behalf of all of the bilateral cross-cousins in attendance. All cousins of the deceased, male and female alike, are entitled to share equally in what is given on these occasions.

Feast items are presented to cousins in a raw state. After receiving it, they will cook and redistribute it to all those who have come to witness the event. Kamea insist that cousins must eat first at mortuary distributions. Once they have been fed, all others in attendance can act as consumers as well. A portion of the food is handed back to the original givers, who will eat a communal meal at the house of the chief sponsor. In this way, a shifting dichotomy is created between feast givers and recipients, one that mirrors the themes of similarity and difference discussed earlier. On the one hand, cross-cousins of the deceased and hosts are sharply distinguished from one another by having each party consume its portion of the comestibles on their own. On the other hand, an image of sameness is emphasized by having feast takers share the food with the original givers. I shall have more to say on this below. Here, the important point to note is that by handing back some of the food to sponsors, cousins demonstrate that their anger is finished; their grief has been consumed along with the food items that have been given to them.

Yet, mortuary feasts do more than mark the termination of the grief of cousins. They also finish the range of interdictions that those closest to the deceased have taken on as a sign of their bereavement. After the final mortuary distributions, all of those who have been avoiding the fruit of the pandanus tree and the food that has been prepared in an earth oven will resume consumption of these items once again. The feeding of cousins, in short, results in the resumption of eating by all others. Mortuary distributions also serve as an occasion upon which those who have grown their hair as a symbol of their grief will have it cut by others who are attending the feast. In the words of one man: "When we cut the hair (*mda tawino*) we remember the dead once again. After the body of the dead has been 'thrown away' (i.e., disposed of) the living get back to their everyday business. When we cut our hair, we think about the dead once again." After the final rites have been held, an officially sanctioned process of forgetting the dead begins.[16]

I noted earlier that Kamea explain the right of cousins to receive things at death as a matter of restitution; cousins deserve to be compensated both for the intensity of their grief and, at least in the past, for their work in preparing the body of the deceased. Yet mortuary exchanges cannot be understood solely in compensatory terms. Implicit in Kamea thought is the underlying idea that the products of cousins are eminently consumable.

First-degree cross-cousins arrange marriages between their offspring. Through the ensuing nuptials, there is a replacement of alternate generations. This process of regeneration is reflected in Kamea kinship terminology. Kamea use the kin terms *ato* (i.e., male ego, FF, MF, DS, SS) and *ati* (i.e., female ego, FM, MM, DD, SD) to refer to males and females in alternate generations. This link is manifest in a number of social practices, which are based on a perceived similarity between the old and the young. For example, although childbirth is seen to be a highly polluting event for men, a woman's father will often tend to this daughter while she is in labor. It is almost as though a perceived link of sameness between alternate generations renders him invulnerable to what would otherwise be a highly contaminating event. One cannot be adversely affected by that which is essentially the same.[17] By "consuming" the children of cousins, one replaces oneself. In mortuary exchanges, there is a corresponding idea that one is entitled to consume the products of one's cousin's labor when he or she dies. One man received pork when his father's sister's daughter died, explained, "My [female] cousin (*nawi*) looked after this pig. I must eat it."

In his description of Tangan mortuary customs, Foster (1995) asserts that the durability of shell items that are given to the matrilineages of the deceased engenders their perdurance as a social unit in the face of death. In describing these exchanges he notes that the qualities of items used in mortuary distributions "are not simply fortuitous features . . . like the color of paper currency but rather properties inseparable from the recognition of the object as a value" (64). It is significant, then, that the items given in Kamea mortuary exchanges lack any enduring qualities. The pigs, garden produce, and game that are given to cousins at death feasts are always given as items to be cooked and consumed immediately, rather than as items with reproductive potential. Pigs, for example, are never presented to cousins as fertile sows, but only as objects to be consumed immediately. For Kamea, I suggest, the items given during funeral distributions are ephemeral (i.e., consumable) because social identities and relationships are also understood to have something of an evanescent quality about them. The items used in Kamea mortuary exchanges symbolize the consumable quality of social relationships—that is, their tendency to undergo perpetual metamorphosis. Marilyn Strathern has remarked that "in a world where social relations are the object of people's dealings with one another, it will follow that social relations can only turn into [other] social relations, and social relations can only stand for [other] social relations" (1988: 172). The exchange items featured in Kamea mortuary distributions are appropriate expressions of this process. Food does not objectify a system of static social divisions, but rather, the processual nature through which social life is created on an ongoing basis.

Kamea constructions of death cannot be interpreted via the logic of our current analytical paradigm. Contemporary accounts of mortuary practices account for the meaningfulness of these rites in terms of their capacity to shore up a system of salient cultural distinctions. Through the activities of human beings, mortuary rites are said to convert the contingency of individual existence into an opportunity for social renewal. Bloch (1982) explains how this process is understood to work:

> The Indonesian, Melanesian, and Chinese opposition between flesh and bones and the different treatment which these should receive seems to be an example of [an] ideological bifurcation. . . . The flesh, the female part, is polluting and has in these cases to be totally dispersed before the bones, the male part, can release the power of fertility and blessing to the next generation. This is the explanation of the temporary burials of Borneo on platforms away from the earth, on which the flesh of the body must first decay before the bones can be buried so that the social order can reproduce itself. This is also the explanation of the common New Guinea practice of cleaning the bones of one's ancestors of any remains of flesh before these can be used to canalise fertility and the power of the clan. (223)

Traditional approaches to death view the body as a corporeal entity that is subject to decay, while the categories of social life are fixed and immutable. Indeed, in Bloch's formulation, it is through the destruction of the former that the regenesis of the latter is achieved. Central to this model is the assumption that what constitutes the individual and society is both self-evident and fixed. Kamea do not share this analytic dichotomy. Individuals and bodies are fully relational from the outset; they do not exist in opposition to a broader social field.

This helps to account for a particular feature of Kamea funerary practices that originally perplexed me while I was staying at Titamnga. Several men and women told me that individual death feasts are not held for a child or an unmarried youth. The final mortuary rites of the very young are always held in conjunction with a feast that commemorates the passing of an adult man or woman. Children, in other words, are never the sole compulsion behind funerary exchanges—they are always honored as an adjunct to secondary identity. Until a boy is initiated and a girl is married, they lack the capacity to enter into social relationships as productive agents. They are still contained within an encompassing maternal identity (see chapters 2 and 3), and under such circumstances, their death does not elicit a singular response. The composite nature of their identity in life is reflected in how they are treated at death.

Throughout the sequence of death rites, cross-cousins are given the role of consumers. In return for their work in preparing the body of the deceased, it is expected that they will eat first at mortuary exchanges. As has been widely reported of other people in the ethnographic literature, Kamea posit a close connection between sex and eating. Indeed, Kamea women are said to be eager to marry because they are "greedy" to consume the juice (*iya coka*) of their husbands-to-be. Through the process of gestation and giving birth, a woman's own body is consumed by her offspring—a process that results in her weakening and eventual death.[18] Men would suffer a similar fate were it not for the *yangwa* tree, the sap of which promotes their greater health and longevity. The cross-cousin relationship evinces a similar concern with consumption, but here it is entirely nonsexual. It is only in the generation immediately following that of cousins that there is a return to the consumption of bodies (through the marriage of their children), rather than things. These bodies are, in turn, defined themselves, in part, through the differentiating activities of cross-cousins in the negotiation of alliance relationships.

It is significant that it is only after the final mortuary distributions have been held that the widow of the deceased is free to marry once again. According to the men and women of Titamnga, when a man dies his wife is placed under stringent taboos to avoid all contact with the cross-cousins of her late husband. Until the finishing feast is held, the widowed spouse is not permitted to "stand in the eye of" (i.e., be seen by) her spouse's cross-cousins. As one man put it: "When my cousin dies, I will get pig and money. This woman—my cousin's wife—she is taboo. She cannot come near me. She is ashamed." After the final mortuary distributions take place, the widowed spouse can resume all interactions with her husband's kinsmen. From this point on, she is permitted to marry and to resume a life with sexual relations. By extending a feast to cross-cousins, the woman is reconstituted as a productive, sexual being who is capable of entering into new relationships.

As the aforementioned indicates, it is not simply the ephemeral qualities of the exchange items presented at death that matters; meaning is also engendered through the paired images of consumption and nonconsumption.

We saw earlier that certain dietary restrictions are associated with death. By their own volition, friends and relatives of the deceased avoid eating pandanus fruit and food that has been prepared by a <u>mumu</u> (stone oven). Cousins, by contrast, are defined largely through their eating behavior. It is noteworthy that they represent the one category of kin who are never expected to take on dietary interdictions at death. Being consumers, cousins turn everyone else into consumers through their activities. As the feast progresses, the distinction between consumers and nonconsumers is progressively mediated. Cousins consume feast items first, and through their work they release all others from their dietary prohibitions. Even feast givers themselves are gradually transformed into consumers through the work of cousins at mortuary exchanges. By the time the feast has ended, the widow of the dead man is free to marry again should she so choose. The role of cousins at death parallels their role in life by turning the unproductive cross-sex sibling relationship into one of productive alliance.

DEATH AND NONESSENTIALIZED STATES

This chapter opened by considering recent debates concerning the significance of cloning in Europe and North America. Since Dolly's birth in 1997, the prospect of human cloning has been the subject of considerable public attention. Indeed, cloning has probably generated a more heated response than any other innovation in the field of biotechnology. In the words of Harold Shapiro (1997), cloning "raises a set of questions about what it means to be human: questions that go to the heart of the way we think about families and relationships between the generations, our concept of individuality and the potential for treating children as objects" (195).

Yet, if Dolly's creation has prompted many of us to reflect on the social and ethical implications of creating life in a laboratory setting, from the perspective of the scientists who created her, her significance lies elsewhere. Until Dolly was born, it was assumed that the nuclei of an adult cell could not be reprogrammed back to an embryonic state. In the early stages of an embryo's development, all of its cells are identical. At this stage, any cell in the developing fetus is capable of developing into any kind of cell in the adult organism. In technical terms, the cells are "totipotent," meaning they are defined by their inherent plasticity and mutability (DiBerardino and McKinnell 1997). As the embryo continues to divide, however, the cells of which it is composed begin to differentiate into vastly different structures— nerves, skin, blood, organs, muscle, and so on. As the cells begin to specialize, they lose their capacity for taking on other functions. Liver cells, for example, cannot become muscle tissue and skin cells cannot be transformed into early embryo cells (Silver 1997: 94). Differentiated cells, or so it was believed, cannot go back in time to a prespecialized state in which their cell functionality was more open-ended (Franklin 2005: 62–63).

By creating Dolly, Ian Wilmut and his team at Roslin Institute demonstrated that this assumption was incorrect. By combining an enucleated egg with the nucleus of an *adult* (i.e., previously differentiated) mammary cell, they were able to create a new animal. "What Dolly shows in principle," claimed Wilmut, "is that we can start again." Wilmut called this process of recapturing embryonic totipotency "de-differentiation." The birth of Dolly demonstrated that you could reverse the genetic clock: "you could make cellular differentiation go backwards" (McLean 2002: 1050). As Sarah Franklin (2001a) has noted, this amounted to a radical suggestion in the scientific literature: "It means that one of the most important formal properties of the biological—the principle of one-way development, and the economy of this principle, that there is an inevitable loss of capacity in the progress towards complexity—has been overturned. The equation of development with specialization, specialization with a one-way temporal scale, and loss of capacity with increased development, all turn out to be technically reversible" (9).

What cloning reveals, then, is the possibility of a recursive temporality. Life need not unfold in a unidirectional sequence. There is the possibility of "doubling back," as it were. The potential for time to "flow backward" contravenes several important premises upon which a genealogical model is based. In particular, it challenges the prevailing Euro-American point of view that kin relations are grounded in an irreversible generational sequence.

Taken for granted European and North American ideas about death articulate with the aforementioned paradigm. Death is seen to represent the cessation of a unique and unrepeatable life. The social relationships that the deceased enjoyed while he or she was alive are understood to end with the passing of the individual. New persons will be born; new relationships forged—but these too will have their own unique, unrepeatable quality. The temporal sequencing that accompanies a genealogical paradigm is distinctly linear and progressive in nature.[19]

For Kamea the unfolding of social forms takes recursiveness as its underlying baseline. As we have seen in previous chapters, what is at issue for the men and women of Titamnga is not the reproduction of individuals per se, but rather the reproduction of sets of relationships. Parents produce siblings, who produce cousins, who produce spouses. The cycle, then, begins anew. Cousins are central to this process of social regenesis. They transform one-blood similitude into gendered states and at the same time, differentiate relationships based on siblingship from those of affinity. This carries with it important implications. If the parent-child tie is assigned the "creative" nexus in Euro-American kinship configurations, for Kamea, by contrast, "reproductive capacity" is distributed among a broad range of different relationships, most of which are nonsexual. Biology does not play the determining role in the elicitation of relatedness. Further, it is not simply birth but also death that acts as an important nexus in the ongoing creation of sociality.

This shift in emphasis significantly influences how we can approach Kamea mortuary practices. What I have demonstrated in this chapter is that Kamea death rites cannot be interpreted via the model that has dominated much of Papua New Guinea ethnography. This model reflects a concern with the ways in which bodily substances given at conception and through postbirth feeding relationships form the basis of a shared identity—one that is then taken apart at death and which furnishes the raw material from which new social identities and relationships are conceived to spring.

From nearly the beginning of our discipline, anthropologists have framed their interpretations of mortuary events in terms of explanations that relate funeral practices to the reproduction of the social order. Central to this approach is the view that the physical body and the social body should be seen as metaphorical equivalents of one another, and that manipulating the former at death (either "symbolically" through exchange activities, or by acting directly upon the corpse) will reconceive the constituent units of which society is composed.

One aim of this chapter has been to expose those assumptions that form the basis of our current theory of mortuary symbolism and to suggest that where sexual reproduction is not used as the principal means of tracking sociality, the meaningfulness of death practices must be sought along alternative lines. It makes little sense to view Kamea death rites as a process of "deconception," when it is not conception per se that forms the basis from which social relationships derive.

The data presented in this chapter raise several additional questions concerning the adequacy of orthodox anthropological theory. To date, most treatments of mortuary practices rest on a firm foundation of Western assumptions concerning the supposed link between gender, sexual reproduction, and sociality. In the ethnographic literature (Chodorow 1974; Ortner 1974; Rosaldo 1974) women have often been associated (either implicitly or explicitly) with "biological reproduction," while men, or so the argument goes, have a closer affinity with the more prestigious task of building broader social relations. Women are equated with nature, men with culture (see MacCormick and Strathern 1980: 348, for a critique of this position). This assumption has been incorporated into our approach to death. Thus, Bloch and Parry (1982) write, mortuary symbolism "identifies women with sexuality and sexuality with death. Victory over death—its conversion into rebirth—is symbolically achieved by a victory over female sexuality and the world of women" (1982: 22).

In Kamea worldview, social relationships and individuated bodies are not perceived as separate domains. Nor is gender understood to be an essentialized state. Among the people with whom I worked, an association between women and the polluting aspects of death is not made. It is cross-cousins and not women who deal with the pollution of death, and they do not frame their activities in terms of a "nature"/"culture" dichotomy.

Kamea mortuary practices are tied to the ongoing creation of social life. But, it is a process that makes sense in their terms, rather than in Euro-American ones. The idea that death allows persons to be taken apart and sent back to the sources from which they came assumes that there is a set of ready-made and enduring social distinctions that exist above and beyond the life and relationships of particular individuals. Selves, in this view, are simply disassembled into their constituent parts (based on a procreative model of how they came together in the first place) and returned to their place of origin—more often than not, to a corporate descent group (Mosko 1983; Munn 1986; A. Weiner 1980). But for Kamea, as we have seen, the distinctions upon which social life is based must be created on an ongoing basis. Gendered states, singular bodies, and differentiated modes of relating are not perceived to be innate but are brought into being through conscious human effort. It makes more sense to see Kamea social life as involving an ongoing process of "production," rather than in "reproductive" terms.

FIVE

Conceiving Global Identities

"VANISHING PEOPLE": THE MAGIC OF THE POSTMODERN ERA

"Racing the clock, two leaders in genetics and evolution are calling for an urgent effort to collect DNA from *rapidly disappearing* indigenous people" (Roberts 1991: 1614; italics added). These words are the first byline in an article by Leslie Roberts, entitled "A Genetic Survey of Vanishing People," which highlights the purported significance of the Human Genome Diversity Project (HGDP). Published in the popular journal *Science,* the paper goes on to describe the seemingly heroic efforts of a dedicated team of researchers (mostly population geneticists and evolutionary biologists) whose intent is to document the "ethnic fingerprints" (R. Lewin 1993: 25) of some five hundred groups of indigenous peoples by sampling their genetic material before it passes out of existence. The study makes use of new techniques in molecular genetics that have become available as a consequence of the Human Genome Project. These procedures allow researchers to isolate minute differences in DNA sequences (known as polymorphisms) that exist not only within populations but between them as well. However, according to Roberts and the teams of scientists whose work he describes, time is running out. Indigenous peoples are "fast disappearing" across the globe, and as they vanish, "they are taking with them a wealth of information buried in their genes about human origins, evolution and diversity" (Roberts 1991: 1614).

Although several recent writings convey a sense of sense of newfound urgency concerning the contemporary plight of Third World people (Associated Press 1996; Cavalli-Sforza 1991; Gillis 1994; Perlman 1993), the idea that indigenous peoples are on the brink of extinction is not new. One is reminded, for example, of Lévi-Strauss's musings in "Tristes Tropiques" (1961), where he expresses a sense of nostalgia for the passing of authentic human differences. Having set out during the 1930s to find a society "reduced to its most basic expression," he is confronted instead only with the "filth" of Western civilization. Shantytowns fill Africa, commercial and military aircraft zoom over South America, tourist shops in Asia stock souvenirs that sport the label "Made in the U.S.A." Travel no longer puts the would-be adventurer face to face with "exotic" places and peoples. Instead, humanity now displays a kind of "monoculture" that has been precipitated by the expansion of Western commodity culture.

And yet, if one seriously considers the message contained in these works, what exactly is on the brink of extinction? When we speak about "disappearing" indigenous peoples, what exactly has (or is) "disappearing"? Certainly, people themselves have not simply "vanished" from the planet—they have not been the unwitting subjects of a conjuring act by a master magician. When Western scholars talk about the "disappearance" of indigenous peoples, what they are lamenting is the loss of a particular idea of what "otherness" entails. What is interesting to note within the present context is that the criterion that is used to define "otherness" has changed substantially over the past few decades. Where Lévi-Strauss once saw "otherness" as entailing divergent sociocultural traditions, today it is increasingly represented in biological terms as a constellation of different gene frequencies. Biology now competes with the notion of culture as the most salient feature of what it means to be a "people."

In this chapter, I pursue this theme in greater detail. More specifically, I examine how biology is coming to be evoked as a social and moral value in emerging forms of global discourse. In earlier chapters, I have argued that a biological paradigm significantly shapes how Europeans and North Americans imagine gender, persons, bodies, kinship, and our relationship to other life forms on the planet. I also suggested that the world looks very different when it is seen through Kamea eyes. In recent years, however, and at an ever increasing pace, a biological framework is being exported as a worldview to other parts of the globe as part of the international flow of power, capital, and meanings.

In this chapter I consider the consequences of globalizing biology. The ensuing discussion will initially take us away from the particulars of Kamea ethnography and to a broader analysis of the implications of extending biology as a conceptual framework to other parts of the globe. I will argue that this development is having an important impact not just on the lives and worlds of non-Western people, but is also precipitating a significant shift in what has long served as the West's core metaphor, the division between "nature" and "culture" (Wagner 1975). To the extent that this distinction has long been central to the Western scientific tradition (including how Europeans and North Americans define what is meant by the "other") it carries with it important implications for the place that anthropology occupies in the academic division of labor. In the last section of this chapter, I return to a discussion of what these developments might mean for a group of people like Kamea. I also reflect upon the broader implications of these developments for the social sciences more generally.

DIVERSIFYING BIOLOGY

In a remarkable book that foreshadowed much of the critical scholarship that has defined the postmodern era, Wagner (1975) outlines the assumptions upon which Western scientific knowledge has developed. Science "advances" to the extent that its practitioners are understood to have learned something "new" about the world. "Discovery" is the name of the game. Yet, more than the pursuit of knowledge is at stake. One of the most compelling points that Wagner makes in *The Invention of Culture* concerns the extent to which Westerners invent themselves and their mode of being in the world through our attempts to rationalize and predict the world of "nature." In a brilliant passage that highlights the relationship between invention and convention, Wagner argues: "Like so many things, our technological culture *must fail* if it is to succeed, for its very failures constitute the thing that it is trying to measure, harness or predict. Science and technology 'produce' our cultural distinction between the innate and the artificial to the extent that they fail to be completely exact or efficient, precipitating an image of the 'unknown' and the uncontrolled natural force. Thus it is that science and technology are aligned on the side of conservatism in modern America" (Wagner 1975: 72; italics added). Science "advances" less from what it *knows* than what it *doesn't know* (and cannot ever possibly know) about the world.

In earlier chapters, I discussed some of the most recent developments that have taken place in the fields of science and technology. Our ability to clone, to produce interspecies life forms, and to create human embryos without the sexual act seems to herald the dawn of a new era. Today, "nature" is increasingly produced in factories and laboratories, and is subject to the supply and demand of an international market. Social scientists have been quick to take an interest in this brave new world (Cannell 1990; Franklin 1997, 1998a, 1998b, 1999, 2001b, 2002; Haraway 1991; Nothnagel 1996; Papagaroufali 1996, 1999; Rowland 1992; M. Strathern 1992a, 1995, 1997). The growing ambiguity between "nature" and "culture," between what counts as "natural" fact and "social" fact, promises to redefine the basis of personhood, perceptions of kinship, property relations, and how we view our connections to other life forms on the planet.

Yet, although we appear to be poised on the brink of a post-Cartesian world—one that promises to transcend the distinction between the world as "given" and the world as "made"—I will argue that the situation is not so simple. In ways that Wagner could have predicted over a quarter of a century ago, the distinction between the "innate" and the "artificial" has not collapsed, it has simply assumed a new guise. What we are witnessing alongside the growth of a biosocial regime is the progressive "culturalization of nature" and the "naturalization of culture" (cf. Rabinow 1996: 241–42).

In what remains of this chapter, I pursue this line of reasoning by examining our current fascination with "biological diversity." Perhaps more than any other term in contemporary Western society, "biodiversity" has emerged as a key symbol of early twentieth-first-century techno-scientific thought (Escobar 1999; Hayden 1998; Zerner 1994).[1] Rarely encountered in popular discourse, even ten years ago, this term has become ubiquitous in both scholarly scientific and nonacademic writing. Encapsulated within its use are a rich array of meanings including: contemporary efforts to stave off species extinction, the growth and proliferation of the conservation movement, a thoroughgoing critique of the destructive tendencies of industrial society, and a growing recognition of the pervasiveness of global interdependencies (cf. Escobar 1998a; Hayden 1998; Zerner 1994).[2] In addition to its signifying the emerging values of the bio-age, *biodiversity* has taken on an operational function. As the term gets put into general circulation, it is performing a rather remarkable feat; it is facilitating the metamorphosis of "nature" into "culture" and "culture" into "nature." This chapter will chart the trajectory of this process, including what it entails with respect to how Europeans and North Americans perceive Third World and indigenous peoples.

My own interest in the contemporary discourse on biological diversity emerged as a direct consequence of my ethnographic research with Kamea. In 1992, shortly after completing my doctoral fieldwork with Kamea, two events took place that immediately captured my attention. Approximately four months after I had returned to North America, I received word that the environmental agency Conservation International (CI)—a group that has its main administrative headquarters in Washington, DC—had moved into the highlands of Gulf Province and founded a bio-research station along the southwest Kamea fringe, approximately three days walk due south of Titamnga. Under the direction of Western-trained botanists and zoologists, they had begun to carry out research on local ecology and to establish a training program for advanced Papua New Guinea students in ecology. CI's plans for the future include creating a "conservation management area" and promoting ecotourism as a non-timber-based mode of economic development in the region.

At approximately the same time that CI was commencing their work in the Kamea region, the Associated Press began to leak word of the so-called Hagahai controversy, involving yet another group of highland New Guinean people. The case, which was first brought to public attention by an Ottawa-based development group, fully emerged in the mid-1990s and began to sharpen debate over what some have termed "genetic colonialism" (Hanley 1996). It appears that in the late 1980s, researchers from National Institute of Health (NIH) in the United States discovered that many Hagahai peoples carried a human T-cell leukemia virus, although they themselves were not afflicted with the disease. Using a blood sample drawn from an unidentified twenty-one-year-old male, NIH researchers established a cell line, a self-perpetuating culture of virus-infected cells. In 1991 they applied for a patent, which was granted three and a half years later (Hanley 1996). The patent document states that the "invention" may be useful in developing a vaccine or devising techniques to test for T-cell leukemia. This means that for the next seventeen years, the U.S. government (or any company that buys the rights to that patent) will have the sole right to use that Hagahai individual's virus-infected cells for commercial purposes.

Initially, these two events struck me as being completely unrelated, apart from the fact that they were taking place in that area of the world where I had carried out ethnographic research. One was focused on saving the planet; the other was concerned with describing microscopic fragments of human DNA. It was only later that I came to realize that these two projects had much in common, inasmuch as they were both geared toward the documentation, preservation, and development of "biological diversity." Yet, while each endeavor was motivated by similar intentions, their repercussions are markedly different. When the notion of biodiversity gets applied to plants and animals, it leads to their progressive "culturalization." When it gets applied to people, by contrast, it leads to their "naturalization." In what follows, I will pursue this point by comparing the aims and intentions of CI with those of the Human Genome Diversity Project.

GLOBAL ECOLOGIES

It was in large part the relative isolation of the Kamea region that attracted the interest of Western environmentalists in the first place. In 1992, working in conjunction with the Foundation for People and Community Development (Papua New Guinea's largest rural development NGO), Conservation International carried out a one-month biological survey of the Lakekamu Basin—an area of land that lies at the confluence of the Gulf, Central, and Morobe Provinces. In addition to Kamea who occupy land that lies along the northern edge of the basin, three other cultural-linguistic groups have also traditionally used resources in the region: the Biaru, the Kurija, and the Kovio (Kirsch 1997: 98) (see map 2). As a consequence of the CI survey, the Lakekamu Basin was assigned a "high priority" rating for conservation efforts on the basis of two criteria: "The biodiversity within the Basin is globally significant. The Basin is exceptional in terms of *numbers of species* (e.g., the richest ant fauna in the world) and is outstanding in terms of *endemism* (New Guinea's biota is essentially unique except for a few species shared with Australia's threatened fragments of remaining rainforest). However, what is most significant in terms of conservation is the size and condition of the rainforest habitat. Within the Lakekamu Basin and adjoining hills lies a vast, intact ecosystem" (Conservation International 1998: 13; italics added).

CI describes its work in the basin in terms of two interrelated aims: to promote the growth of scientific knowledge (particularly concerning the "discovery" and documentation of new species), and to create a link between an increase in the socioeconomic well-being of landowners and the conservation of natural resources (a point to which I return below). With

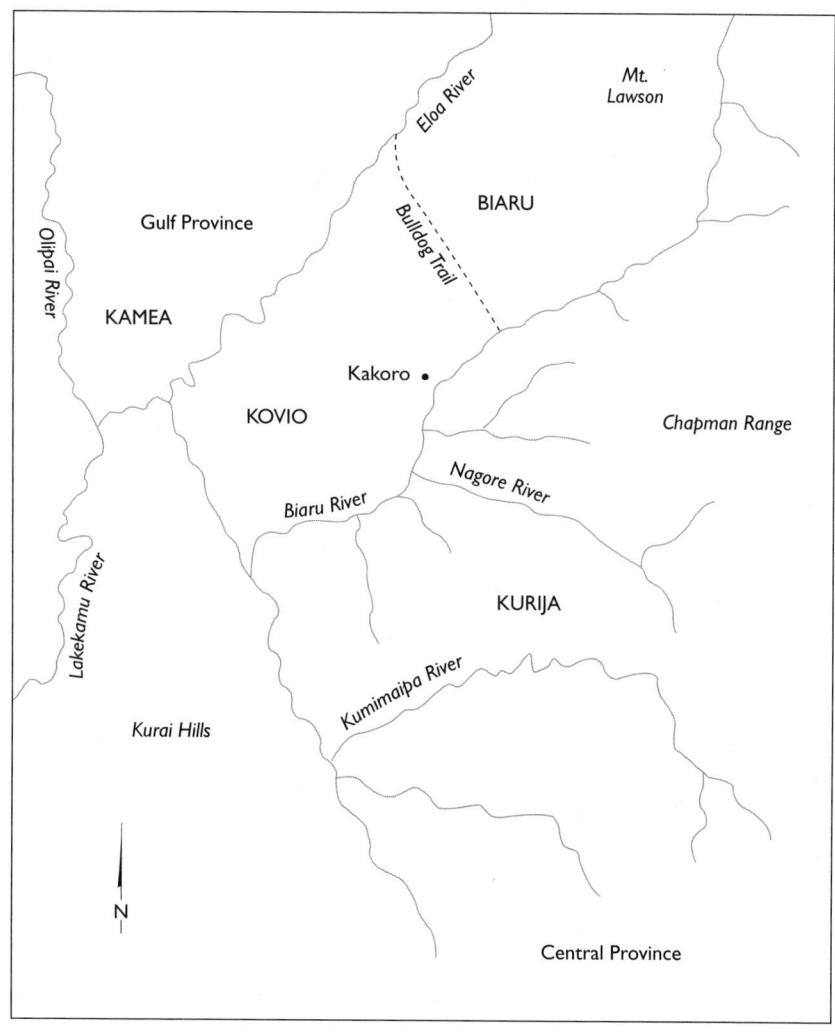

Map 2. The Lakekamu River Basin region. Adapted from Kirsch 1997: 99.

these aims in mind, CI established a research station in the fall of 1993 and shortly thereafter assembled a team of botanists, zoologists, and ecologists with the intent of surveying the basin's resources.[3] Through a combination of research- and development-related initiatives, CI hopes that viable alternatives to mining and logging will be in place before such industries enter the region and pose a threat to the ecological balance of the basin.

Like most environmental movements, the philosophy of CI is underwritten by the premise that the "natural world," and consequently the social order that it sustains, faces imminent destruction owing to our overuse of nonrenewable resources. The rapid rise of human population and the tremendous growth in technologies have so changed man's place in the ecological balance, it is argued, that what is called for is nothing short of a fundamental change in our attitudes toward "nature" and life in general (Herscovici 1985: 17). Ultimately at risk, we are told, is "the air we breathe, the water we drink, the soils and seas that feed us and the living creatures that give us fibres, medicines and countless other products" (Conservation International 1997a). Because countries like Papua New Guinea have yet to be overexploited, they are seen to be prime candidates for Western conservation measures. In addition to their work in Papua New Guinea, CI sponsors programs in other countries, including Bolivia, Botswana, Colombia, Costa Rica, Guatemala, Madagascar, Mexico, and Peru, to name but a few. Funding for these ventures is derived from a combination of private donations and corporate sponsorship, including generous endowments from the Ford Motor Company, Anheuser-Busch, Bank of America, Intel, McDonald's, Subway restaurants, Starbucks Coffee, and the Walt Disney Company.

The self-proclaimed "mission" of CI, as revealed in their published literature, is to "protect the earth's biologically richest areas and to help the people who live there improve the quality of their life" (Conservation International 1997a). To achieve these aims, CI strives to identify what they call "hotspots": areas that contain the highest concentration of plant and animal diversity and appear to be under imminent threat of destruction.[4] The value of biodiversity is seen largely in economic terms. "Without biodiversity," they write, "human beings could not survive. . . . Wild species are the source of all of our foods as well as many of our most important fibers, building materials and medicines" (Conservation International 1997a). CI cites an impressive array of statistics in support of their claims: approximately 40 billion dollars are spent annually on plant-derived drugs. Another 10 billion is generated each year through the sale of minor rain forest products, including rattan, nuts, potpourri, bamboo, and the like.

In their efforts to halt the tide of destruction that "contemporary society" has unleashed, CI advocates a program of "sustainable development."[5] Perceiving their activities as "one step ahead of chain saws" (Conservation International 1997b), they attempt to reach NGOs and local communities before large-scale extractive industries cause irreversible damage to natural habitats. The business ventures promoted by CI reflect this emphasis on sustainability. More than simply generating a cash income for local communities, they are designed to impress upon local peoples that their on-going participation in the cash economy (indeed, in some cases, their very inclusion within it) depends on their adopting long-term conservation measures.[6] Thus, in addition to offering an online ecotourism adventure travel service (which provides consumers with information on "safe," "eco-logically-friendly" "exotic destinations" (Conservation International 1997a), CI also manages the Web-based Rainforest Marketplace, which sells sus-tainably harvested rain forest products. CI's "flagship enterprise" in the world of business was the Tagua Initiative, a project that was undertaken in South America in 1990. CI describes this venture: "Instead of selling their land to loggers and cattle ranchers or *burning it for agriculture,* rainforest communities from Ecuador and Colombia now earn a living by harvesting tagua palm nuts, which they sell to button and jewelry manufacturers. In a few years, this mini-industry has generated more than two million dol-lars. Through enterprises like these, CI has found international markets for other rainforest products including allspice, Brazil nuts, cocoa, potpourri, natural dyes, and oils used in cosmetics" (Conservation International 1997a; italics added). In the New Guinea highlands, CI's list of proposed activities include initiating small-scale logging ventures, galip nut busi-nesses, bioprospecting, and adventure tourism.

Business metaphors abound in the published literature of CI. When I first began to peruse their documents on the Web, the first entry I turned to concluded with the statement that "resource protection is good business for everyone" (Conservation International 1997a). As I continued to browse their Web pages, I became acquainted with the earth's "biological wealth," its "ecological riches," and the need to adopt an "economic approach to sav-ing the planet" (Conservation International 1997a). The commodification of the nonhuman world that is attendant in this approach is evinced in what CI describes as being one of its most "innovative approaches to conservation" (Conservation International 1997a). In 1987 they launched their first-ever "debt-for-nature-swap." Using funds donated by the Connecticut-based Weeden Foundation, they purchased the face amount of U.S. $650,000 in Bolivian bank debt for U.S. $100,000. In exchange for extinguishing their financial obligations, the government of Bolivia agreed to set aside some 2.7 million acres of land for "conservation purposes." Since that time, nearly 7 million dollars has been exchanged in eighteen different transactions leveraging over 15 million dollars for conservation ef-forts (Conservation International 1997a).[7] In CI's own words, the success of this project was based on their perception that "the debt crisis would actually be a window of opportunity for conservation organizations" (Conservation International 1997a).[8]

Although CI's programs are modeled on a system of economic relations that has its origins in the West, it perceives the world in distinctly global terms. Using "nature" as a point of departure from which to critique late capitalist culture (cf. Wagner 1975), it claims to take a stand not only against the excesses of corporate industry, but also against colonialism and its attendant history of exploiting indigenous people. In the past, they say, multinational corporations have profited immensely from their use of rain forest materials, but have given little back to tropical countries in return. For example, the discovery that Hodgkin's disease and several forms of childhood leukemia can be treated with drugs derived from the rosy periwinkle, a plant endemic to Madagascar, has meant little to the people of that country. Multinational corporations have profited immensely from this discovery while local peoples have received nothing in the way of recompense.

CI promotes a radically different vision of the world. Employing the rhetoric of a liberal, democratic ideology, they argue for an extension of "rights" to all citizens of the world. Basing their stand on the view that rain forest countries are entitled to benefit from the development and sale of their endemic species, CI has been active in the development of international policy as it pertains to the ownership of genetic resources (Conservation International 1997a). One of its most aggressive campaigns to date has centered on the field of "bioprospecting"—the development of genetic material derived from plant and animal species (Conservation International 1997a). In a pamphlet that describes their "shaman's apprentice program" (Conservation International 1997c), alternatively described as "the hunt for genetic resources," CI states that they "encourage local tribes to record their knowledge, be proud of their culture, and profit from it economically" (Conservation International 1997c).[9] Saving the planet, it would seem, can (and should) be a rewarding experience for everyone.

AN EXPANDING NATURAL MARKET

While couched in its own particular brand of rhetoric, the discourse of Conservation International is fairly representative of contemporary biodiversity campaigns. Construed at one level as a redemptive act, an attempt to protect "uncontaminated" zones of the nonhuman world as "pristine" and "undisturbed,"[10] the narratives produced by biodiversity advocates often read more like a manual on saving contemporary capitalist society. As Charles Zerner puts it: "Nature is analogized to a warehouse, a library, or a safe-deposit box containing fixed, valuable, and threatened commercial assets" (1994: 72). Previously "uncapitalized" parts of nature, including the very genes of living species, become imminent sources of commercial value, awaiting only the power of science and technology to release their profit-making potential (cf. Escobar 1996; R. Lewin 1993). Critics of contemporary environmentalism have been quick to point out the contradictions inherent in this movement.[11] Arturo Escobar has argued, for example, that far from challenging the basic premises of modern industrial society, campaigns to conserve biological diversity represent a *deepening* of capitalist interests in the Third World. In a 1996 article, Escobar notes, for example, that in contemporary discussions, "the key to the survival of the rainforest is seen as lying in the genes of the species, the usefulness of which [can] be released for [monetary gain] through genetic engineering and biotechnology in the production of commercially valuable products, such as pharmaceuticals. Capital thus develops a conservationist tendency, significantly different from its usual reckless, destructive form" (Escobar 1996: 47).

Yet, if the rhetoric on biodiversity is fairly uniform in treating "nature" as one big shopping mall, it evinces a certain amount of confusion in knowing exactly how to situate indigenous peoples in the dialogue. On the one hand, indigenous peoples appear at first glance to be assigned a favored position in the emerging rhetoric. They are given the role of "stewards" in charge of preserving the last remaining vestiges of biodiversity on the planet. Yet, if this appears to place them in a position of empowerment, it is a position that is nonetheless riddled with contradictions. For as we shall see, contemporary discourse not only constructs the "native" as "super-hero"; it also has the effect of "essentializing," "homogenizing," and ultimately "naturalizing" those very people upon whom the survival of the planet supposedly depends (Brosius 1997).

Because conservationists currently concentrate the bulk of their efforts on safeguarding and protecting tropical rain forest environments, it is crucial that they present their campaigns for public consumption as being in synch with the interests of local peoples. A common rhetorical tactic in the literature is to present rain forest dwellers as "natural allies" of environmentalists (cf. Ellen 1986). Activists often achieve this impression by presenting local people as having a "sacred" connection with the land (Brosius 1997: 54). In the words of Conservation International, for example: "A mutual interest in conserving and wisely using the land and resources creates a strong alliance between conservationists and indigenous peoples. . . . [T]o work for the conservation of biological diversity in the world's hotspots is to work with indigenous peoples who have more than an economic attachment to the ecosystem where they live. They also have ancestral, spiritual, even linguistic ties that are rarely shared by loggers, ranchers and slash-and-burn colonists who also seek to occupy tropical lands" (Conservation International 1997a).[12] Although well intentioned, this commentary on the sacred carries with it a number of important consequences (cf. Brosius 1997). The most obvious is that it rides rough-shod over nuances of local meanings in an attempt to present Western audiences with a compelling sound bite of Tribal Wisdom. While Kamea, as we have seen, certainly *do* perceive their connection with the land in what can be described as "social" and/or "personified" terms (see chapter 1), it would be erroneous to extend their worldview to other Papua New Guineans, let alone to rain forest peoples who occupy other parts of the globe. As Brosius (1997) has argued, environmentalist rhetoric renders generic the very diversity that it seeks to represent.

Perhaps even more surprisingly, couched within contemporary narratives is an attempt to reconcile the perspective(s) of modern industrial society with the aims and intentions of the "rest of the world." It is implicitly assumed that indigenous peoples the world over possess a desire to be incorporated into the world capitalist economy. Brazilians really "want" to manufacture potpourri, just like highland New Guineans secretly covet owning and operating their own bed-and-breakfast establishments. Indigenous peoples are portrayed as sharing the same attachment to a capitalist economy as the West. In a somewhat messianic vein, they are presented not only as would-be laborers in this system, but also as fledgling "scientists" who can solve the problems of our current environmentalist crisis. In the words of activist Alan Duning, indigenous peoples "possess in their ecological knowledge an asset of incalculable value: a map to the biological diversity of the earth on which all of life depends. *Encoded in indigenous languages, customs and practices may be as much understanding of nature as is stored in the libraries of modern science*" (quoted in Brosius 1997: 54; italics added). Yet, if the entire world emerges as being one big happy family, equally committed to the values of science, capitalism, and conservation, human diversity is not completely erased. Instead, it assumes a "naturalized" form.

When I left Papua New Guinea in the spring of 1992, the search for genetic "resources" was deepening and taking on new dimensions. Occurring in tandem with the patenting of Hagahai cell lines was the birth of an ambitious plan called the Human Genome Diversity Project. Conceived initially in 1991 by Stanford University population geneticist Luigi Luca Cavalli-Sforza and deceased evolutionary biologist Allan Wilson, the aim of the HGDP has been to "create a data base of human genetic variation" before this diversity disappears from the planet (Hayden 1998: 174). The plan entails collecting blood and tissue samples from literally hundreds of indigenous groups world-wide (R. Lewin 1993: 25). Targeted groups are selected on the basis of their geographic isolation and their presumed ability to answer questions of compelling interest to social scientists. Researchers hope to use the data generated from this project to cast light on the evolutionary history of human beings, to reconstruct "human genealogy," and when possible, to analyze these samples for clues concerning human disease susceptibility and immunity (Cavalli-Sforza 1991; Gillis 1994; Hayden 1998; R. Lewin 1993).

As described by the major players involved, the HGDP was undertaken in an effort to "broaden the vision of the human gene pool that undergirds the Human Genome Project—a three billion dollar international effort to create a single reference map of the genetic code of the human species" (Hayden 1998: 179). Up until now, researchers of the human genome have focused on the similarities among all of us. Scientists believe that 99.9 percent of the genes contained within any person precisely match those of his or her neighbor. The remaining 0.1 percent of one's genetic makeup varies, however, and it is these variations that are of interest to Cavali-Sforza and his associates (R. Lewin 1993; Marks 1995). Arguing that there is no such thing as *the* human genome, members of the HGDP have asserted that their own efforts are "a necessary complement to the larger and better funded Human Genome Project" (Hayden 1998: 179). Indeed, according to Sir Walter Bodmer (president of the Human Genome Organization, or HUGO) the new survey is a "*cultural obligation* of the genome project" (quoted in Roberts 1991: 1615; italics added). A statement issued in 1994 explains the work of the HGDP committee in the following way: "Without this project, science will largely define 'the' human genome, with its historical and medical implications as that carried by a small number of individuals of European ancestry. . . . At a time when we are increasingly concerned with preserving information about the diversity of many species with which we share the Earth, surely we cannot ignore the diversity of our own species" (Human Genome Diversity Project, quoted in Hayden 1998: 179).

There is a perceived urgency in the work of Cavalli-Sforza and associates. Located in the bodies of so-called "isolated" populations is an invaluable resource—a "pure" gene pool, supposedly engendered through years of endogamy and reproductive isolation. Indeed, "it is specifically as repositories of unmixed DNA that 'isolated' groups become marked as ancient or ancestral populations" (Hayden 1998: 180, cf. Roberts 1991: 1614). Yet, this very important resource is in danger of imminent destruction. "Remote" and "ancient" people are becoming "extinct" with each passing day, and with them, a unique opportunity for scientific study. According to one HGDP member: "Of the roughly 5000 languages in the world, 90% are expected to be lost or doomed to extinction by the twenty-first century. Genetically distinct populations could disappear with them, some by physical extinction, *but mostly by admixture with other groups*" (Gillis, quoted in Hayden 1998: 180; italics in original). Members of the HGDP hope to avert this tragedy. Collected samples of DNA are to be stored in the American Type Culture Collection in Rockville, Maryland, where they will be entered into a permanent database of human genetic variation. In this way, even if indigenous people disappear, their cells can be studied by scientists in perpetuity.

ON THE NATURE OF "OTHERNESS"

A profound irony exists with respect to how indigenous people are conceptualized and represented in contemporary biodiversity campaigns. When advocates of conservation speak of the "threat" that is currently faced by indigenous peoples, they are not referring to a perceived loss of sociocultural traditions. Indeed, the aim of this campaign is to incorporate Third and Fourth World peoples more fully (and under carefully scrutinized conditions) into the scope of global capitalism. Contemporary narratives are grounded in the assumption that people the world over share similar hopes and dreams, not to mention ideas concerning the "proper" use of "nature." Where indigenous peoples *do* differ from Europeans and North Americans, however, is in their genetic makeup. As isolated pools of genetic material, they warrant our concern and immediate attention. "Tribal" peoples must be sampled and studied before their gene pool is contaminated through "interbreeding" with "outsiders."[13] As Hayden succinctly puts it, contemporary biodiversity campaigns applied to indigenous peoples introduce the somewhat unnerving notion of "death by reproduction" (1998: 181).

I have argued that an important corollary of contemporary conservation rhetoric is that it serves to "naturalize" peoples of the Third and Fourth World. Indigenous peoples are presented as an endangered "subspecies" of sorts, not unlike the trademark Panda that appears in the upper left-hand corner of World Wildlife Fund paraphernalia. The methodology that has come to guide the Human Genome Diversity Project makes such an equation nearly inevitable. When the project first got underway, its principle leaders—Cavalli-Sforza and Wilson—fell out over the sampling strategy. Lewin describes the point of contention: "Cavalli-Sforza wanted to sample aboriginal populations that had long been isolated and were culturally and linguistically discrete. Wilson, by contrast, pushed for a geographic-grid strategy, where sampling would simply take place every 50 or 100 miles, regardless of who was there" (R. Lewin 1993: 27). In the end, Cavalli-Sforza's approach won out, and genetic diversity was defined as a product of one's cultural heritage.

The Web pages of Conservation International are replete with colorful examples in which indigenous peoples are implicitly likened to rain forest species. CI's 2000 home page, for instance, opened with a picture of a highland New Guinea male, resplendent in face paint and traditional feathered headdress, as he takes his place alongside (and presumably on equal footing with) a starfish, flower, tree, lizard, and nonhuman primate. The similarities between the Human Genome Diversity Project and the Human Relations Area File (HRAF) of anthropology's own history are worthy of comment. Both are projects of cataloguing (rather than studying) diversity and both are fueled by the fear of diversity's imminent extinction.[14]

Conservationists' efforts to draw an equation between indigenous peoples and the world of "nature" are hardly new. Indeed, the "naturalization" of indigenous peoples has a long-standing history in Euro-American thought (see, for example, Lutz and Collins 1993, on *National Geographic*). Yet, it is important to note that although indigenous peoples have long been equated with "nature" in Western thinking, the basis along which such a connection is made has changed radically over the past few years. Until recently, this association was established on the basis of cultural differences. Third and Fourth World peoples were seen to be "more natural" than Euro-Americans because they lacked the external trappings of "civilization." Their relationship to the world of "nature" was seen to be "more direct," less mediated by cultural artifacts than that of Euro-Americans. Here, it is "culture" (often defined in a quantitative sense, in the sense of having more things) that serves to differentiate indigenous peoples from Euro-Americans. Human beings are said to be essentially "the same" the world over in terms of their biological makeup; where we differ from one another is in terms of our cultural traditions. Biodiversity campaigns have the effect of turning this proposition on its head. It is no longer possible to speak with confidence about a "universal human nature"; indeed, our so-called "natural differences" are increasingly called upon to assert cultural boundaries.

In addition to "naturalizing" indigenous peoples in unprecedented ways, modern conservation discourse has the paradoxical effect of "naturalizing" a capitalist economy. This trend is a relatively recent outgrowth of the conservation movement. Twenty years ago capitalism and environmental ethics were seen to be contradictions in terms; the former was asserted to exist at the expense of the latter. Today's conservation rhetoric promises not only to achieve a rapprochement between these terms—it makes the very survival of tropical ecosystems dependent on the globalization of a capitalist economy. This point of view undergirds what CI describes as their "new" model for conservation, a plan that entails in their own words "environmental entrepreneurship": "In this model, international businesses form partnerships with local groups who sustainably harvest rainforest products. In so doing, they create incentives for local people to save forests rather than cutting them down" (Conservation International 1997a). Capitalism emerges here not as a threat to endangered ecosystems, but as an important vehicle in promoting a "greener" world.

THE POLITICS OF A "NEW" BIOLOGY

In a 1997 text on kinship, Robert Parkin reflects on the significance of recent developments in biotechnology. In a book that weighs in at 208 pages, only 3 pages are devoted to a discussion of recent advances in biotechnology. Following a disclaimer that the separation between sexual reproduction and parenthood is "nothing new" in the ethnographic literature, Parkin justifies giving this topic short shrift by stating: "At present . . . the new reproductive technologies are a 'problem' for the essentially Western societies that have developed them rather than for other societies in the world" (126).

As I hope to have illustrated in this chapter, such a view is not entirely accurate. The preceding discussion has focused on biodiversity conservation and the Human Genome Diversity Project. These issues were selected because they have had a direct bearing on the site of my own ethnographic research. But a biological worldview is increasingly finding a home in non-Western countries—through an ever-expanding range of venues. Genetically modified organisms now support much of the so-called developing world. In vitro fertilization is used in India, Singapore, China, Africa, Israel, Kuwait, and Saudi Arabia. The World Health Organization sponsors family planning workshops throughout much of the Pacific (Sykes 2006). Contraceptives, in the form of condoms, the pill, and injections, are widely distributed throughout the south (Bledsoe 2002). Innovations in biotechnology are having a marked impact on more than how kinship configurations are defined. As Sarah Franklin has noted in a discussion of Parkin's text, "The oversight in [his] account . . . is its inability to appreciate the ways in which kinship in the context of new reproductive technologies does not concern merely 'new ways of making babies' but a much wider set of issues, such as how knowledge is produced, how capital is accumulated, and how identity categories are transformed" (Franklin 2001b: 319; cf. Ginsburg and Rapp 1991).

By means of conclusion, I return to the issues raised by Wagner in *The Invention of Culture* (1975). In this book, Wagner demonstrates the extent to which Western perceptions of "self" and "other" are entangled with Euro-American efforts to manage what stands as a key symbolic operator in the West—a clear-cut distinction between the world of "nature" and the world of "culture." As Wagner demonstrates in his work, managing this distinction is no laughing matter; preserving the terms of this dialectic is coterminous with safeguarding an entire way of life, particularly a certain way of knowing the world.

The advent of the "bio-age" appears, at first glance, to threaten the West's core metaphor of creativity. Today, scientists do more than simply rationalize the world of nature; one could say that, within the contemporary world, nature is "made to order" (Franklin 1997; Rabinow 1996; M. Strathern 1992a). The expanding use of new reproductive technologies, genetic engineering, prosthetics, and cloning, to name but a few of the most astonishing advances, have allowed us to become habituated to the dizzying pace of scientific discoveries. The ability of science to impress us with its seemingly impossible feats has become "extraordinarily ordinary" (Hayden 1998: 191) over the past several years.

Yet if our very "successes" threaten to relativize the basis upon which past achievements were made, we have responded to this potential "crisis" through a further (and predictable) act of innovation: we have reversed the terms of the dialectic. As "nature" increasingly takes on the guise of "culture," "culture" is undergoing a process of "naturalization." In the emerging rhetoric of postmodern, techno-scientific thought, Third World and indigenous peoples are increasingly represented to Western audiences as being "culturally" similar to ourselves; where they differ is in terms of their underlying "nature."

The ethical implications of this brave new world warrant a certain amount of introspection. In chapter 1, I examined Kamea human-environmental relations. We saw that the resources upon which Kamea depend for a living, the land and the different species of flora and fauna that they utilize, are not simply appropriated via preexisting social ties, but instead furnish a venue through which salient social distinctions are created in the first place. Gender and different categories of social relationships emerge from the differential uses to which the nonhuman world is put. The men and women of Titamnga do not simply act upon (or manage) the nonhuman world; rather, the latter is directly implicated in the ongoing creation of social identities and relationships. The world as it is imagined and acted upon by Kamea is very different from that which impels contemporary environmentalist campaigns.

In recent years, Kamea (like other indigenous peoples) have become the unwitting recipients of global environmental planning. Through the initiatives of conservationists, they are learning about the need to preserve our "genetic resources," and that it is incumbent upon everyone, the world over, to limit our consumption of nonhuman species. Yet, as we have already seen, the specter of "uncontrolled consumption" that motivates contemporary environmentalist discourses has little place within the context of Kamea environmental relationships. For here, it is not consuming *too much* that jeopardizes the fabric of social life, but rather consuming *too little,* thereby alienating one from the land and ultimately from other people. Recall the plight of Netsap discussed in chapter 1. Having been disenfranchised from the system of land tenure, Netsap lost more than the means of production; he was effectively cast outside of the orbit of positive sociality.

If consumption is perceived in negative terms by Western environmentalists, for Kamea, by contrast, it represents the highest ideal, in that it furnishes the means by which humans come to define themselves as moral beings, fully enmeshed within the scope of positive sociability. We are perhaps not premature in speaking about an emerging form of biological imperialism—one that will have significant repercussions in the years to come.

One of the most important points to emerge in Wagner's *The Invention of Culture* concerns the moral implications of Western scientific thought, particularly with respect to how Euro-Americans draw upon scientific rhetoric as a means of framing and legitimizing their encounters with non-Western peoples. In his work, Wagner asks us to take seriously the implications of our own inventiveness, what it means when Europeans and North Americans insist upon using our own metaphors as a means of understanding and representing others. He demonstrates the ethical quandaries involved when he recounts the story of Ishi, the last surviving Yahi Indian. Housed in a museum until his death, "Kroeber and others would take Ishi back into the hills so that he could demonstrate Yahi techniques of bushcraft" (Wagner 1975: 28). When Ishi lived in the Anthropology Museum of the University of California (on Parnassus Heights in San Francisco), from 1911 to 1916, he was asked to document his "otherness" by engaging in acts of "culture." In the contemporary world of global capitalism and biosocial modes of relating, demonstrating "otherness" has taken on a new meaning. Instead of taking people off to the bush and asking them to demonstrate their cultural expertise, we immortalize the cell lines of "endangered" peoples in the American Type Culture Collection in Rockville, Maryland. The gene has become a dominant symbol of contemporary social life (Nelkin and Lindee 1995).

Conclusion
Conceptual Displacements

October 22, 1991. Titamnga Village, Papua New Guinea

Bipahu, Patura, Wilimal, and I are seated around the slowly ebbing fire that burns in my <u>haus kuk</u> (kitchen). The evening is cool. We sip tea to ward off the evening chill and contemplate an issue that has been troubling Wilimal for the past few weeks. Wilimal is in the process of planning a feast that he intends to present to his in-laws in three days' time. The intended recipients of the feast are the mother and father of his first wife, Itipaney, a woman to whom he has been married for approximately six years. Wilimal frets, as he has been doing for some time now, over what he should give them. Pig, sweet potatoes, garden produce, and wild <u>caruka</u> nuts are all planned constituents of the presentation—a substantial gift by Titamnga standards. Still, he wonders whether this will be enough. Perhaps he should defer the feast until he can acquire a second pig, or has saved up enough cash to purchase some store-bought rice and tinned fish? Wilimal also wonders if his efforts will finally pay off. He has been making presentations to Itipaney's parents for as long as he can remember—indeed, long before he married their daughter—but to no avail. Even after all of this hard work, Itipaney has failed to conceive a child. His second wife, Mobi, has already borne a child named Libi, who is now two years old. Itipaney, however, remains frustratingly barren. Wilimal wonders just how many more presentations he will need to make before Itipaney is able to carry a child. His ongoing efforts on his wife's behalf are proving to be a continuing drain on the family's resources.

I began this book by considering the Del-Zio case, which made headline news during the mid-1970s. Like Itipaney and Wilimal, Doris and John Del-Zio faced issues of infertility. However, the discrepancy between their cases could not be more striking. Working in partnership with the biomedical establishment, the Del-Zios attempted to correct "Doris's problem" by circumventing her blocked fallopian tubes to achieve pregnancy. Through science and technology, egg and sperm were brought together in a laboratory setting, albeit with failed results. Although Itipaney and Wilimal also faced difficulty conceiving, the situation they encountered was radically different. Instead of confronting "failed biology," the situation they were dealing with involved a "failure" in social relations; namely, Wilimal's inability to meet his in-laws' expectations. The solution Wilimal turned to—gift giving/exchange—was intended to address the problem.

Since the discipline of evolutionary biology first emerged in the mid-nineteenth century, it has served as an orienting framework for Europeans and North Americans. It has figured centrally in how Western audiences approach a diverse array of social configurations, including what goes under the guise of "kinship," "gender," "bodies," "persons," "ethnicity," and the relationship between human beings and other life forms.

Three ideas have been central to a biological framework, particularly to how this system of ideas has filtered into the popular imagination. First is the notion that all the constituents of the human and nonhuman world can be plotted on a single genealogical diagram. By virtue of the vivifying mechanism of "descent with modification," the organic world has come to be characterized by unity and diversity at the same time. Second is the associated idea that sexual reproduction fuels the motor of evolutionary change insomuch as it defines not only species boundaries, but also generates the physical variation upon which natural selection acts. Third is the notion that organisms adapt to their environment. The environment is, in turn, viewed as a set of external constraints (cf. Ingold 1992).

This book has presented a series of snapshots concerning how the world looks when this system of ideas is held in abeyance. In particular, I have drawn upon two contrasting conceptual frameworks—innovations in biotechnology and Kamea ethnography—to provide a set of destabilizing devices through which to view the assumptions that accompany a biological paradigm.

Recent developments in science and technology have captured the popular imagination by virtue of their ability to upset traditional assumptions. Today, organisms can be created in the laboratory and in the absence of sexual intercourse. Recombinant genetic technology (rDNA) has made it possible to swap genetic material not only between species, but also between kingdoms. Cloning allows for replication, thereby challenging the view that reproduction necessarily entails the production of unending newness. As Franklin, Lury, and Stacey (2000) have pointed out, the end result of biotechnology has been a respatialization and retemporalization of genealogy: "The neat genealogical system that Darwin described is no longer closed, tree-like, or unified" (190). What we are witnessing through these technologies is a reformulation of life itself.

Recently, many scholars have turned their attention to the social and ethical implications of what it means to create life—both human and nonhuman—independent of a process of physiological reproduction (Andrews 1999b; Davis-Floyd and Dumit 1998; Edwards 2000; Franklin et al. 1993, 1997; S. Kahn 2000; Ragoné 1994; Rapp 1999; Silver 1997; M. Strathern 1992a, 1992b; Wilmut, Campbell, and Tudge 2000). These works are important because they reveal culturally specific Western assumptions about relatedness that have long informed Euro-American understandings of kinship (Holy 1996), not to mention anthropological interpretations of non-Western societies. For example, when one considers high-profile controversies about surrogate motherhood, posthumous reproduction, and "mix-ups" at fertility clinics, it becomes difficult to discern what biology grounds (or does not ground) and to specify what kinds of ties should flow from reproductive acts (see Franklin et al. 1993).

A central contention motivating this book is that however enlightening science and technology studies have been, the challenge they offer is partial. These analyses provide an "internal" critique of Euro-American assumptions in that they use the "new biology" to rethink the categories upon which the "old biology" was founded. New forms of baby making are used to interrogate old forms of baby making (Weston 2001), but the focus on biology remains intact. It is my belief that a deeper and more nuanced understanding of Euro-American worldviews may be had by placing these discussions into dialogue with concepts from a different part of the globe. In this book, I have undertaken this task by engaging Kamea ethnography in the conversation. The end result has been the creation of a three-way dialogue, in which the "new" biology challenges the "old" and Kamea perceptions challenge both of these.

Kamea represent a particularly apt foil against which to view those assumptions that inform Western cultural logic by virtue of their tendency to ground sociality in processes other than biological reproduction. They provide us with a telling glimpse into a world in which sexual intercourse is not seen as the basis from which social life springs. Although the people with whom I worked are more than able to articulate where babies come from, conception is not used as a means of tracing social relationships through time. Indeed, as we have seen, Kamea are rather blasé when it comes to the entire subject of sexual reproduction, viewing it as an unremarkable (and certainly, an unremarked upon) feature of human relatedness. Because neither a mother nor a father shares substance in common with their offspring, the ideology of "shared characteristics" that underwrites conventional notions of descent fails to have the same explanatory power there that it does in European and North American societies. To understand what connects and disconnects people in their world it is necessary to adopt a conceptual framework that is radically different from that which informs biology.

Kamea social life is based on a contrast between male and female relational configurations. Men and women each engender their own modes of relating that can be viewed in terms of a contrast between lineal and lateral relationships. Relationships defined through women are of a short temporal duration and receive their most potent expression in the horizontal ties of "one-blood" siblings. Kamea men, by contrast, are implicated in the definition of relationships through time. While of fundamental importance to Kamea social life, these modes of relating are not innate. The intentional separation of male and female that takes place at marriage and through the activities of the men's cult begins a process of differentiating "one-blood" siblings that leads to the elicitation of gendered capabilities. While the capacity to act as a male or female has its roots in the struggle of substance that takes place in a woman's womb prior to birth, this is only the beginning. What is laid down as a *potential* must be activated through concrete human effort if it is to be brought to fruition. In my analysis of Kamea, it has been necessary to focus on those processes through which salient social distinctions are brought into being, rather than assume that they are given in "the nature of things."

The creation of those differences upon which male and female adult reproductive competence is based begins to take shape early on as bridewealth is paid on a woman's behalf. These payments do not compensate an agnatic group for the loss of a daughter's or sister's fertility at marriage; instead, they are used to create that capacity. By virtue of these presentations, the "containing" capacity of the bride-to-be is brought into being. If femaleness is created by evoking a woman's "containing" capacity, maleness rests on the opposite move: an act of decontaining that is brought about through the activities of the men's cult. For Kamea, gendered capabilities (including the capacity to form certain kinds of relations) must be brought into being; they are not inherent in the materiality of bodies themselves.

One important implication of the emergent and nonessentialized nature of Kamea sociality concerns the differing temporal implications of male and female relational forms. Men draw upon the land and other resources to inscribe their identities in the nonhuman world. This gives relationships defined through men an enduring temporal quality that relationships defined through women lack. Indeed, as we have seen, the history of men's relationships to the land is what gets remembered through time and forms the basis of intergenerational continuity, rather than genealogical connections.

This association of men with an image of permanence fits well with Kamea ideas concerning the durability of male and female bodies. Kamea assert that men and women destroy themselves in the act of reproduction. A woman's strength is progressively drained through each child that she bears, until finally she succumbs to her wasted condition and dies. Men, I was told, face a similar predicament with respect to semen depletion and the ongoing demands that women and children place on their sperm. Engaging in sexual relations and the production of offspring means embarking upon a road to premature aging and death. There is, however, one important difference that needs to be underscored. Men have at their disposal the rejuvenating sap of the *yangwa* tree, which they can drink to supplement their declining reserves of strength. Masculine bodies, like the relationships they form through time, evince a kind of permanence that is wholly unavailable to Kamea women. Just as a woman's body is consumed by the children that she bears, so, too, her place in the system of social relations is brief—confined to the definition of intragenerational relations.

Throughout this work, I have used Kamea understandings to address three interrelated issues. First, I have been concerned with revealing the extent to which a great deal of theorizing in the social sciences implicitly takes a reproductive model as its underlying baseline. I have argued that in a conceptual universe where biology is not understood to form the underlying basis of kin connections, many tenets of traditional anthropological theory are called into question. Kamea mortuary practices, for example, cannot be interpreted via the standard model that has dominated much of the ethnographic literature, a model that holds that substances given at birth and supplemented through postbirth feeding relationships form the basis of a shared identity, which is then taken apart at death, furnishing the "raw material" from which new social identities are conceived (Bloch and Parry 1982; Mosko 1983; A. Weiner 1980). It makes no sense to see Kamea death rites as a process of "deconception," when it is not conception per se that forms the basis from which social relationships are perceived to spring. Similarly, in a world where bodies and persons are not defined in terms of their essential autonomy and self-determination, it is necessary to seek new meanings for the roles of taboo and reciprocity in social life. Among Kamea, prolonged feeding behavior (i.e., giving a young girl game to eat on a continuing basis) may precipitate salient social distinctions, while food prohibitions can reveal an underlying state of unity. Through these and related explorations, I have highlighted the extent to which a biological model permeates Euro-American social science thought. David Schneider (1984) once cautioned Western-trained anthropologists not to assume that their own understandings of kinship were salient in the societies they were studying. I have suggested in this book that not taking the "facts of life" for granted in our analyses of non-Western societies will entail more than overhauling how we approach procreation and conception.

Second, this book demonstrates the utility of considering "kinship studies" in relation to anthropological investigations of the environment. Although a concern with human-environmental relations has been central to the discipline for decades, it has tended to be approached in a very specific way. Uniting several schools of thought, including Marxist (Marx 1976; Marx and Engles 1970), functionalist (Damas 1969; Hardesty 1977; Lee and DeVore 1968; Vayda 1969), structuralist (Lévi-Strauss 1966, 1967), and symbolic interpretations (Bird-David 1993; Descola 1992; Descola and Pálsson 1996; Durkheim and Mauss 1963), is the view that the organic world is somehow "external" to human social life. One can adapt to the environment, pass it on as a form of heritable property, model interpretative schemes after it, or impress a prefigured model of society upon it, but it is always "other" in relation to human beings and their activities. In this book we have seen that such a framework cannot be sustained with respect to Kamea. Kamea sociality emerges in tandem with the kinds of relationships that people form with the nonhuman world (cf. Ingold 1990, 1991, 1993, 1994, 2000; J. Leach 2003).

The third issue I have considered in this work concerns the implications of globalizing a biological worldview. Through recent developments, including the growing call for biodiversity conservation, a biological model is increasingly finding a home among non-Western audiences. This is bringing in its wake important processes of social change that are likely to continue, if not expand, in years to come. One consequence of this development is that it is precipitating a shift in terms of how Europeans and North Americans view non-Western peoples. Increasingly, indigenous and Fourth World peoples are being filtered back to Western audiences as being "just like us" in terms of their appreciation of a capitalist worldview. They are seen as different in terms of isolated pools of gene frequencies. To the extent that social scientists have considered the repercussions of this issue, they have tended to focus on the political and economic implications. GMOs, we are told, will strengthen already existing forms of dependency between the north and the south (Shiva 2000). Patenting Hagahai cell lines is tantamount to treating non-Western "others" as a form of heritable property (Pottage 1998). While these critiques are certainly well placed, they capture only part of the problem. They reflect concerns that are paramount to Europeans and North Americans; namely, the idea that power and money go hand in hand and often motivate people's actions. It is important to remember, however, that Kamea notions of gender, bodies, persons, landscape, and social relations will all be implicated should Conservation International achieve their intended aims in the Lakekamu Basin. We are dealing with an entire cosmology of life that cannot be summed up through discussions of "global bio-politics."

As will be evident, the insights of David Schneider (1968, 1984) have served as a source of inspiration throughout this work. In particular, I have been guided by his maxim that anthropologists should reflect critically upon the presumptions that enter into their own vision(s) of the world. What often goes as "science," and is "purified" through its appeal to so-called "objectivity" (Latour 1993; Wagner 1975) may be steeped in a plethora of unexamined assumptions.

I have used in this book a comparative framework to reflect upon the implications of a biological worldview. Comparative studies have had a long-standing, if somewhat tortuous, history in anthropology. Once the defining hallmark of ethnographic research, cross-cultural comparison fell into disrepute during the second half of the twentieth century, along with attempts to "solve" what had previously been defined as universal problems of the human condition. However, recently, comparative analyses have witnessed something of a resurgence, particularly with respect to anthropological investigations of kinship. In a work published in 2004, for example,

Janet Carsten talks about the possibility of reinvigorating kinship as a cross-cultural domain of research. She argues that the recent "collapse" of any hard and fast distinction in the West between "nature" and "culture"—a collapse precipitated in large measure through the rise of biotechnology—has been liberating. In particular, "it puts the West into the same analytical frame as non-Western cultures . . . [insomuch as we can no longer assume] that . . . in the West what is social and what is biological are firmly and clearly separated in opposed domains, [while] in non-Western cultures they are inextricably mixed up" (Carsten 2004: 189). The "blending" of what was once viewed as the radically separate domains of social versus biological kinship in the West supposedly places Euro-American social configurations on the same analytical footing with those found in other parts of the world. Like can finally be compared with like.

The position that I have taken in this book is different on two scores. First, I would argue that it is not accurate to say that Kamea "blend" or "mix up" the so-called social versus biological aspects of kinship. A biological framework has never had relevance for Kamea—they perceive the world through distinctly nonbiological eyes. Hence, the challenge that has recently confronted European and North American audiences concerning the dislodging of a long-standing conceptual framework has little in common with Kamea perceptions of the world. The fact that Westerners currently fret about cloning and parent-child gamete donation stems from the challenge these developments pose to a previously accepted way of viewing the world. As I have argued throughout this book, Kamea do not share this interpretative framework. A world in which biology has come to be unmoored is very different from one in which it never existed in the first place.

Second, I would question whether it is accurate to claim that the domains of the "social" and the "biological" have really collapsed in the West. It seems to me that we have not witnessed the disappearance of these domains, but rather a world in which they are assuming different and often unpredictable guises. Marilyn Strathern (1992a) has argued that the postmodern era is characterized by an ethos in which there is simultaneously *more* nature and *more* culture. In recent years, Europeans and North Americans have witnessed an exaggerated emphasis on the biological idiom. The first "test-tube" baby was born in 1978. In 1998—a mere twenty years later—approximately eighty-two thousand IVF procedures were carried out in the United States alone (Henig 2004: 233). The growing demand for this technique indicates the extent to which Europeans and North Americans continue to draw upon (and in ever increasing numbers) a biological framework as the underlying basis of the kin connections. At the same time, however, biology is no longer defined as existing apart from human artifice. It is shaped and molded in IVF clinics, in the boardrooms of Monsanto, in courtrooms that debate the meaning of bodily integrity, and in the bodies of sheep and pigs that are custom-made to manufacture human proteins. Today, the phrase *reproductively viable offspring* often connotes the productive merging of genes and financial capital. We live in a world in which "human culture" is increasingly defined on the basis of gene frequencies, and where "human nature" is increasingly understood in terms of capitalist wants and desires (cf. Rabinow 1992). Perhaps even more importantly, we live in a world where there is an ever deepening need to reflect upon the common, and not so common, ground of the human condition.

NOTES

INTRODUCTION

1. *In vitro* is Latin for "in glass."

2. The case did not actually go to trial until July 1978, the same month in which Louise Brown—the world's first test-tube baby—was born in Oldham, England. Commentators generally agree that had the case gone to court in 1974, the year it was filed, the Del-Zios would have had little chance of success. "The whole process of IVF seemed like something out of a science fiction story, so it would have been difficult to convince a jury that the Del-Zios had, indeed, lost a realistic chance to become parents" (Andrews 1999b: 17). In 1978, however, the jury was influenced by developments occurring on the other side of the Atlantic.

3. See also Kass 1971, 1972, and 1989; McEnroe 2000; and Wolinsky 1988, for accounts that predict similarly dire consequences for the then-emerging reproductive technologies.

4. In an interesting discussion of a similar point, Nelkin and Lindee link today's metaphoric imagery of the gene to the medieval Christian conception of the immortal soul: "The gene has become a way to talk about the boundaries of personhood, the nature of immortality, and the sacred meaning of life in ways that parallel theological narratives." It follows, then, that manipulating DNA "becomes a sacrilege, a violation of sacred ground" (Nelkin and Lindee 1995: 41).

5. It is, of course, ironic that Rivers felt that the greatest advantage of his method was that it would preserve indigenous culture against the onslaught of European influences. He writes: "It is almost impossible at the present time to

find a people whose culture, beliefs and practices are not suffering from the effects of European influence, an influence which has been especially active during the last fifty years. To my mind, the greatest merit of the genealogical method is that it often takes us back to a time before this influence had reached the people. It may give us records of marriage and descent and other features of social organization one hundred and fifty years ago, while events a century old may be obtained in abundance in all of the communities with whom I have myself worked, and I believe that with proper care they could be obtained from nearly every people" (Rivers 1968: 109).

6. This being said, there are some indications that Kamea conceptions are not unique. As Marilyn Strathern (1988: 235) has pointed out, Trobriand Islanders share the view that neither a mother nor a father has any type of substance-based connection with their offspring. The same, apparently, holds true with respect to the Etoro of Papua New Guinea and the Reite (J. Leach 2003; M. Strathern 2001: 234).

7. In an influential essay, Ginsburg and Rapp (1991) take up a similar point with respect to the globalization of a biomedical model of reproduction. More specifically, they argue that the spread of medical hegemony oftentimes displaces or competes with indigenous practices, and may lead to the disorganization of local forms of knowledge (318).

8. In the West, we think that conception occurs in an instant, through the one-time meeting of a sperm and an egg. Many other people see conception as a far more lengthy process, often entailing several acts of sexual intercourse.

9. The term *Kamea* is an administrative designation only and goes wholly unrecognized by the local people themselves. Government officers use this term to refer to those speakers of the Kapau language who reside in Gulf Province and are administered from the district office at Kaintiba. Another twenty-three thousand Kapau speakers reside across the provincial border in Morobe Province, where they are administered from Menyamya and Aseki.

10. More recently, several other ethnographic studies have also been undertaken of Angan peoples. Mimica (1981, 1988, 1991) conducted research with Yagwoia speakers during the late 1970s, while Bonnemère (1993, 1996) and Lemonnier (1991) studied Ankave peoples during the late 1980s. For further information on Angans see also Blackwood (1939a, 1939b, 1940, 1950, 1978), Fischer (1968), McCarthy (1964), Simpson (1953), and Sinclair (1961).

11. Underlined terms are those given in Neo-Melanesian (<u>tok pision</u>), italicized terms are those from the Kapau language.

1. CULTURAL LANDSCAPES

1. In addition to speaking out against GMOs, Bove has launched protests against American junk food, U.S. trade tariffs, globalization, and the World

Trade Organization (Jamieson 1999; Nichols 1999; Sancton 1999). In 1999, he captured the media spotlight when he tore down a half-built McDonalds near his home in Nillau (Jamieson 1999).

2. Bove has been likened to Poland's Lech Walesa (Sancton 1999) and a latter-day Robin Hood, championing the fight of traditional French agriculture and cuisine against the globalized onslaught of fast food (Graham 2001). He has been praised by France's highest officials, including President Jacques Chirac and Prime Minister Lionel Jospin (Daley 1999).

3. *Bacillus thuringiensis* is a naturally occurring soil bacterium that produces the organic toxin Bt. Organic farmers in the United States have used it to control insects that feed on their plants.

4. According to several commentators (Falkner 1999; T. Rubin 1999) Europe's greater resistance to GM crops is related, at least in part, to a crisis of confidence in the industrial production of food and the role of biotechnology in particular. This can be traced back to previous food scares revolving around bovine spongiform encephalopathy (BSE)—"mad cow disease"—or salmonella-infected eggs. If this is indeed the case, one can expect the burgeoning negative reaction in North America to widen.

5. As will be discussed in chapter 5, a remarkably similar set of concerns underlies the Human Genome Diversity Project. Proponents of the HGDP insist that there is an urgency in collecting genetic information from the "last remaining pockets of indigenous peoples" before their gene pool is sullied through interbreeding with outsiders. As Cori Hayden (1998) has eloquently put it, this introduces the somewhat paradoxical notion of death by reproduction.

6. Such was believed to be the case in the much publicized account of the monarch butterfly. In 1999, researchers at Cornell University performed a laboratory experiment. Dusting Bt corn pollen on plants populated by monarch butterfly caterpillars, they found that many of the caterpillars died (Nash 2000). Subsequent studies have suggested that genetically modified Bt corn poses no risk to human health or the environment, including the monarch butterfly, and that the chances of a caterpillar outside of laboratory conditions encountering the amount of pollen used in the Cornell study are virtually zero (Niles 2001).

7. As we shall see in the pages that follow, my wording here is entirely inappropriate. Since Kamea do not perceive the world in terms of a set of essentialized distinctions, the notion of "crossing" such boundaries makes no sense. I employ this terminology only as a means of drawing readers' attention to an important contrast that exists between Kamea and Euro-American views of the world. I return to this point in the final section of this chapter.

8. Although Descola's argument bears some resemblance to that set forth by Durkheim and Mauss (1963), his emphasis is different. Instead of focusing upon symbolic logic (i.e., the classification of nature), Descola is concerned with the

domain of practice—how human beings structure their social relationships with the nonhuman world.

9. I am borrowing here the terms set out in Descola's (1992) original essay. As will become evident, a nature/culture dichotomy has little salience in terms of how Kamea perceive and act upon the world. One of the goals of this chapter is to show how social life is lived in a world where such a distinction has little recognized significance (cf. Goodale 1980; MacCormack and Strathern 1980).

10. My requests to accompany women on their trips to gardens were decidedly anomalous in the face of this more solitary work ethic.

11. This does not seem to be a new pattern. Beatrice Blackwood (1978) reported similar findings when she worked with Kapau speakers during the early 1930s.

12. I tried repeatedly to elicit some kind of exegetical statement that would help to elucidate the basis upon which male and female crops are distinguished. This line of inquiry met with little success. I did learn that most (perhaps even all) of those crops described as male are planted in pipia (rubbish) piles—that is, they are surrounded by the organic debris of other cultigens. Chapters 2 and 3 will tease out an indigenous association between women and an image of containment. It may be that crops become male when they issue forth from an encompassing female source.

13. This point was completely missed by early Australian patrol officers. When they first began to make sporadic incursions into the Kapau/Kamea area, they consistently encountered a plethora of names, both spatial and social, that seemed to bear no obvious relationship to one another. Local people were frequently described by patrol officers in terms of their "nomadic tendencies" (Blackwood 1939a; Hennelly 1912; Higginson 1908; McCarthy 1964; Sinclair 1961; Zimmer 1969), and "movement between hamlets" was a frequently reported complaint in patrol reports. But rather than recognize such mobility as part and parcel of the indigenous system, colonial administrators assumed that it represented a "breakdown" of the "clan system" and a corresponding dissolution of the traditional land tenure system.

2. INSUBSTANTIAL IDENTITIES

1. I follow the press in labeling the twins as black. For further discussion of the case see Anonymous 1995; Dalton 2002; Rozenberg and Stokes 2002; Smith and Wright 2002; Verkaik 2002.

2. The theme of biosocial reductivism in the academy has recently been taken up in a lively collection of essays entitled *Complexities: Beyond Nature and Nurture* (2005), edited by Susan McKinnon and Sydel Silverman. Spanning all of the four major subfields of anthropology, the authors highlight a growing tendency

in the social sciences to account for social life almost exclusively in biological terms.

3. This point has been discussed at length by M. Strathern (1992a, 1992b, 1997).

4. The first patient received her own embryos, but they were of a poorer quality and failed to develop into a pregnancy.

5. The most obvious ethical questions include: Does the right to reproduce—or not reproduce—survive death? Can genetic resources be inherited like other forms of property? If parents cannot force their living children to give them grandchildren, is it ethical and appropriate for them to acquire such rights after death? See Andrews (1999a, 1999b) for a detailed discussion of the ethical implications entailed.

6. While eggs are more susceptible to anoxia than sperm, and hence more difficult to collect, the retrieval and storage of ova has also become relatively commonplace in recent years (Soules 1999).

7. During an interview with the media, Jean Garber is quoted as saying: "I don't care what people think. Some people say that I am trying to duplicate Julie. . . . Well, I'd give anything if I could duplicate her" (Hiscock 1997).

8. As we shall see in this chapter and the ones that follow, such a conclusion would be nonsensical from the perspective of Kamea.

9. This point is taken up in detail in an enlightening article by C. Cussins (1998), in which she discusses, among other things, the perceived implications of having a sister serve as a surrogate for her brother's child.

10. In the dominant Euro-American model, even if one chose not to "recognize" this relationship in social terms, the "physical" connection would nonetheless remain. Biology creates a bond that cannot be broken, unlike "social ties," which can be severed.

11. See also Wagner (1967, 1977), A. Strathern (1973), J. Weiner (1982), Counts and Counts (1983), and LiPuma (1988) for earlier formulations of a similar model.

12. These are my terms, not those of Kamea. However, as will become evident, my use of them is in keeping with indigenous understandings of sociality and how it unfolds.

13. As we have seen, this same idea underlies the use of NRTs. Residents of Alltown, as described by Edwards (forthcoming) repeatedly claimed that siblings were the most suitable donors of gametes in cases of infertility because they were interchangeable in a genetic sense.

14. A fairly common occurrence given that polygyny is frequent.

15. Kamea, it would appear, are not unique in their tendency to assert that parents do not share an embodied connection to their offspring. In *The Gender of the Gift* Marilyn Strathern (1988) points out that Trobriand women are not connected to their children via a bond of substance (231–40). She also notes in

a more recent publication that among the Etoro of PNG, generations are not perceived in continuous terms—a fundamental tenet of genealogical thinking (M. Strathern 2001: 229–35).

16. This refers collectively to both bridewealth and childprice payments.

17. This represents an important contrast to Euro-American procreation ideas, according to which a child is related equally and *in the same way* to both the mother and the father.

18. Virtually all highland systems have a patrilateral bias in terms of how intergenerational social relationships are tracked through time.

19. Andrew Strathern (1971a) and Marilyn Strathern (1987) have noted similar points with respect to the Wiru. It would be fair to say that Kamea more closely resemble the Wiru than they do the majority of those highland people upon whom this model of matrilineal payments has been based.

20. During my two-and-a-half-year stay at Titamnga, virtually all of the marriages that took place in the village were in keeping with the aforementioned pattern. Of the older existing marriages, I recorded only two cases where the spouse did not belong to the appropriate category, and in both instances, the persons involved in the match were quick to point out that they had married atypically.

21. In addition to making gifts, most youths attempt to ingratiate themselves with the parents of their future brides by offering assistance in a range of household tasks, including building fences, making gardens, collecting firewood, and so forth.

22. Like bridewealth, these payments are typically made to a girl's mother and her collateral kin. I will have more to say about this later.

23. In this, Kamea are radically different from other Angan people. Godelier (1986: 20–25) notes, for example, that the preferred mode of marriage among the neighboring Baruya people is sister exchange. When, however, this proves to be impossible, it is acceptable for bridewealth to be paid. Ideally, however, a daughter should then be sent back in the following generation. See, also, Herdt (1981) on this point.

24. Bulmer's (1967) classic piece, "Why Is the Cassowary Not a Bird?" also highlights the marked place that cousins hold in Karam thought. Here too, cousins occupy the conceptual space between siblings and affines, a place that is likewise characterized by simultaneous closeness and distance.

3. EMBODIMENTS OF DETACHMENT

1. See Anonymous 1986; Balisby 1987; Bonavoglia 1987; Cassens 1987a, 1987b; Gallagher 1987; McNamara 1989; and Pollitt 1990, for further details concerning the Stewart case.

2. Judge Mac Amos based his decision on the fact that Elias had charged Stewart under an inappropriate statute. He argued that the section of the penal

code under which Stewart had been indicted was not intended to penalize women for their conduct during pregnancy but rather to enforce child support arrangements. The inclusion of fetuses within the statute's definition of "child" was aimed only at husbands who abandon their pregnant wives and refuse to pay for their share of pregnancy costs (Anonymous 1988: 994). Judge Amos went on to recommend that the state pass a "more appropriate" bill that would "protect the life of the unborn child under certain narrowly defined conditions" (Bonavoglia 1987). Shortly thereafter, a bill was brought before the California legislature that extended existing child-endangerment laws to viable fetuses (Gallagher 1987: 45).

3. The expanding use of posthumous reproduction, as discussed in the last chapter, is so controversial precisely because it goes against this principle.

4. Lévi-Strauss (1967) noted a similar point early on when he argued that taboos on eating clan totems mark the unity of clan members.

5. One man likened the smell of these animals to human feces.

6. Here, I follow Lambek (1992), who writes: "If the semantic content of the taboo elaborates who or what one is not, it is the practice of the taboo that substantiates who one is" (248).

7. Indeed, the tabooed food items should not even be cooked together in the same container, as other food in that the smell of the interdicted game would spoil what was otherwise an acceptable meal.

8. To anticipate the argument that follows, he is considered to be "female-like" but at the same time, he lacks a woman's containing capacity. This places him, perhaps more accurately, in a position of undetermined gender.

9. In the New Guinea highlands, this has been seen to be a particularly important mandate, given the climate of endemic warfare that characterized precolonial socialities. The men's cult has been seen to promote an exaggerated sense of male solidarity, a necessary condition for the creation of an effective fighting force.

10. Indeed, it could be argued that the respective eating habits of mother and son perpetuate this containment. See Lambek (1992) on the performative quality of taboo.

11. The warrior function has, of course, lapsed since the government put an end to fighting but the promotion of "strength" (*yannganga*) continues to be a primary aim of the initiation sequence (see Bamford 2006, for further details).

12. In large part, this reflects my own decision; however, it was also influenced by the circumstances of my fieldwork. Kamea openly debated whether or not I should remain with women during the penultimate moments of the marita ceremony. Some men actively campaigned for my presence in the men's cult house, arguing that the importance of this ceremony was something that the "outside world" should understand. Others, however, asserted that the presence of a

woman would ruin the efficacy of the rites. Because almost nothing is known about the significance of women during Angan initiations, I expressed a preference for remaining with the mothers and sisters of initiates. This brought the discussion of my participation to an abrupt end.

13. Menstrual blood and birth fluids are seen to be particularly dangerous to men.

14. See Bamford (2006) for a broader discussion concerning the significance of shifts in initiation practices.

15. Men may or may not have an accurate perception here. Certainly, no woman ever admitted to me that she had handled the bullroarers. At the same time, however, this may be a "publicly acknowledged" secret, known by all but admitted by none.

16. Official male dogma maintains that women are totally ignorant of the very existence of the bullroarers. Based on my conversations with many women I do not believe this to be the case. Many women with whom I spoke knew what the *mautwa* were, but claimed they were bound, particularly in the presence of men, to profess ignorance concerning its existence.

17. I asked several of my Kamea consultants whether a woman's holding on to the bullroarers was seen to contribute to their power or efficacy. Everyone with whom I spoke answered no; women simply hold on to them so that they can be removed later.

4. (IM)MORTAL UNDERTAKINGS

1. Since the birth of Dolly, successful cloning using this technique has been reported in mice, goats, pigs, and horses.

2. A key element that enabled the Roslin group to perform this feat with a nonembryonic cell was inducing the donor cells to enter a phase of quiescence by reducing the level of nutrients that are normally used to support their growth in a culture—a process that evidently makes it easier to reprogram the already differentiated adult cells (Marwick 1997: 1103; Wilmut, Campbell, and Tudge 2000: 200–201; Wilmut et al. 1997: 813).

3. Dolly died in February 2003. By the age of six, she had developed arthritis in her left hip and knee, and her DNA exhibited signs of premature aging. In particular, her telomeres—structures that cap the end of chromosomes and become progressively shorter with age (Cibelli et al. 2002)—were more typical of a much older animal. Since Dolly's birth, several researchers in the fields of zoology and cell biology have expressed safety related concerns when it comes to cloning. Studies have indicated a live birth rate of less than 5 percent, and have shown that survivors suffer from a high rate of deformity and disability (Holden and Kaiser 2002: 601; Jaenish and Wilmut 2001: 2552; McLean 2002: 1054; Pickrell 2001: 2061). In addition, cloned fetuses seem to grow unusually large in

utero, posing a potential risk to not only the fetus but to the gestational mother (McLean 2002: 1054).

4. Ironically, in some discussions, cloning is imagined to produce precisely the opposite effect. In a passage that brings to mind Lévi-Strauss's (1969) argument concerning the origin of the incest taboo, Rao (2002) asserts that cloning is dangerous because it "frees individuals from the need to connect with others and engage in marriage or any kind of intimate relationship in order to have children. In so doing, *cloning could be viewed as radically individualistic* and ultimately antisocial or even alienating, the paradigm right of isolated individuals" (1009; italics added). As this quote indicates, anticloning rhetoric leads to the paradoxical position of having "too much" and "too little" individualism at the same time. The cloned child is said to suffer by having his or her right to a unique identity usurped. At the same time, the person who donated the cell is brought to task on the grounds of excessive autonomy.

5. This point is taken up at length by Marilyn Strathern in *After Nature: English Kinship in the Late Twentieth Century* (1992). See, in particular chapter 1.

6. Most deaths are attributed to sorcery. In fact, people claim that sorcery is more common today than it was in the past, an unanticipated by-product of colonial rule and the loss of warfare as an accepted means of settling disputes.

7. Children might take only a few weeks to dry, but adults took considerably longer.

8. I asked several people at Titamnga if this was done in an effort to incorporate the spirit (*hikwapa*) of the dead, but was told repeatedly that only shamans had the capacity to join with spirits of the dead.

9. Mimica (1991), who worked with the neighboring Ikwaye people, documents an interesting contrast to the practice of Kamea. There, sisters were expected to directly consume the cadaverous fluids of their brothers.

10. Kapau is the language of Kamea. Mbaginta (1976) has written about the related funeral practices of Simbari-Angans to the north of Kamea. Mimica (1991) has described the treatment of corpses among the Ikwaye.

11. This is a Kapau term indicating sorrow.

12. In some parts of the Kapau/Kamea area, the dead were stowed in trees, rather than in rock shelters. The men and women of Titamnga are certainly aware of this practice, but claim to have put their dead in caves only.

13. Significantly, the persons who remembered acting as their cousin's undertakers in the past report the smell of the corpse as one of their most salient memories of the experience. Metaphorically, one could say that death becomes the point at which the "nose" (i.e., one's cousin) inhales itself.

14. The gendered implications of this are interesting to consider. Siblings, the product of female sociality, at death come to be stored in a stone tomb—an element of the landscape, which as we have already seen is constitutive of male sociality.

15. Within the contemporary setting, an interesting inversion to this practice is emerging. People now often place cups, plates, knives, and spoons (i.e., commonly owned everyday items) that once belonged to the dead alongside the *he'aka* in the trees. In the past, the body of the corpse was preserved while the memorial to the dead was allowed to decompose. Today, the reverse is taking place. With Christian burial, Kamea realize that it is now bodies (rather than personal effects) that are more likely to decay. Steel cups and plates are more permanent than traditional mourning ornaments.

16. Except as we have already seen in chapter 1, the dead are not fully forgotten. In the case of men, at least, their legacy of working the land will likely be remembered through time.

17. Persons in alternate generations also share in common a number of taboos, particularly those concerning the use of space, which marks them as a singular category in numerous contexts.

18. In a personal communication, Susan McKinnon has pointed out the irony of this situation to me: women act as the containers of their offspring, but reproduction also engenders a situation where offspring in some sense also come to "contain" their mothers.

19. As noted in chapter 2, the very thought of reversing the anticipated directionality of relationships fills Europeans and North Americans with intense discomfort.

5. CONCEIVING GLOBAL IDENTITIES

1. As the term is currently used, *biodiversity* refers to the sum total of life forms that exist on the planet. The Biodiversity Treaty defines biological diversity as "the variability among living organisms from all sources including inter alia, terrestrial, marine and other aquatic ecosystems and the ecological complexes of which they are a part; this includes diversity within species, between species and of ecosystems" (Convention on Biological Diversity, quoted in Eisner and Beiring 1994: 95).

2. If one looks at the history of environmentalist discourse over the last couple of decades, it is possible to note that an important shift in rhetoric has taken place. In the mid-1980s, most campaigns couched their agendas in a discourse of "rights." Like people, nonhuman species were said to have "rights," and therefore could not be carelessly used and abused by human beings. Contemporary campaigns seem to have abandoned the discourse on "rights" in favor of a celebration of biodiversity.

3. Funding for this venture was provided by a USAID grant-in-aid.

4. Interestingly, the extent to which these criteria apply to the Kamea region seems to be ambiguous. CI points out that the basin seems to be "fairly free of exportable timber species," and that the soils are "unsuitable for plantation agri-

culture" (CI 1998: 9). Minor gold reserves have been located near the Olipai River. However, given their small scale, it is predicted that mining operations "would have very *limited impact on the natural environment* (Filer and Iamo 1989, quoted in Kirsch 1997: 103; emphasis original).

5. By "contemporary society," CI means modern, industrial, Western society, which they see as a global phenomenon, or on the verge of becoming so.

6. In this, CI resembles other mainstream environmental groups that have adopted "third wave" (Luke 1997: 32) environmental policies (see Joyce 1992; Raven 1990; Tangley 1990). Here, collaborative links are formed with state and capital, rather than pitching environmental concerns that challenge them (see Escobar 1995, 1996, 1998b, and 1999, for a particularly good discussion of this point).

7. Debt-for-nature swaps are now in place with the governments of Costa Rica, Ghana, Guatemala, Madagascar, and Mexico.

8. Unfortunately, CI fails to mention what happens to the indigenous peoples who presumably used the land before it was turned into conservation areas.

9. Issues concerning the recognition of intellectual property rights as they pertain to indigenous peoples have proven to be among the most troublesome and vexing for contemporary environmentalists.

10. Space prevents me from pursuing this point in greater detail, but the idea that Third World countries are somehow "untouched," "natural," or lacking in human intervention is, of course, grounded in an important set of Western assumptions concerning the meaning of "nature" (see Cronon 1995; Ingold 1993).

11. Joyce refers to environmentalism as "ecocapitalism" (1992: 399); Luke calls it "ecocolonialism" (1997: 31). Similar critiques have been levied by other scholars (see, for example, Ferguson 1994; Zerner 1994, 2000).

12. Note that in this passage, slash-and-burn agriculture is seen to come from the "outside." Indigenous peoples are assumed to be hunters and gatherers. Like other rain forest species, indigenous people collect, rather than produce, their means of subsistence. This point has been taken up in an eloquent discussion by Ingold (1994).

13. This, paradoxically, is seen to be a more pressing concern today than it was in the past, given the spread of global capitalism.

14. I am grateful to Andrew Walsh (personal communication) for drawing my attention to this point.

REFERENCES

AEBC (Agriculture and Environment Biotechnology Commission)
    2003   *GM Nation? The Findings of the Public Debate.* London: Agriculture and Environment Biotechnology Commission.

Allen, R.
    2002   Scandal of Mix-ups at IVF Clinic: Anguish for Women Given Wrong Embryos. *Evening Standard* (London), Oct. 28.

Andrews, L.
    1992   Surrogacy Wars. *California Lawyer* 12 (10): 42–49.
    1999a   The Sperminator. *New York Times,* Mar. 28.
    1999b   *The Clone Age: Adventures in the New World of Reproductive Technology.* New York: Henry Holt.

Anonymous.
    1986   Fetal Abuse? Against Doctor's Orders. *Time,* Oct. 13, 8.
    1988   Maternal Rights and Fetal Wrongs: The Case against the Criminalization of "Fetal Abuse." *Harvard Law Review* 101 (5): 994–1012.
    1995   Black and White Twins Born after Test Tube Mix-up. *Jet,* July 24.
    1998   Eco Soundings. *Guardian,* Dec. 2.
    1999   You're Carrying Someone Else's Child. *Newsweek,* Apr. 12, 61.
    2000   Japanese Granddads Become Sperm Donors: Preserving the Bloodline Important to Families. *Florida Times-Union* (Jacksonville), Nov. 19.

2002 Parents in Sperm Mix-up Are Still Childless. *Western Mail* (Cardiff, UK), Nov. 5.

2003 Black Twins to Stay with Couple after IVF Mix-up. *Irish Times,* Feb. 27.

Associated Press

1996 World's Indigenous Groups Rally against Genome Diversity Project. *Daily Progress* (Charlottesville, VA), June 12.

Balisby, S.

1987 Maternal Substance Abuse: The Need to Provide Legal Protection for the Fetus. *Southern California Law Review* 60 (4): 1209–38.

Bamford, S.

1997 The Containment of Gender: Embodied Sociality among a South Angan People. Ph.D. diss., University of Virginia.

1998a Humanized Landscapes, Embodied Worlds: Land and the Construction of Intergenerational Continuity among the Kamea of Papua New Guinea. *Social Analysis* 42 (3): 28–54.

1998b To Eat for Another: Taboo and the Elicitation of Bodily Form among the Kamea of Papua New Guinea. In *Bodies and Persons: Comparative Perspectives from Africa and Melanesia,* ed. M. Lambek and A. Strathern, 158–71. Cambridge: Cambridge University Press.

2004 Embodiments of Detachment: Engendering Agency in the Highlands of Papua New Guinea. In *Women as Unseen Characters: Male Ritual in Papua New Guinea,* ed. P. Bonnemère, 34–56. Philadelphia: University of Pennsylvania Press.

2006 Unholy Noses. In *Embodying Modernity and Postmodernity: Ritual, Praxis, and Social Change in Melanesia,* ed. S. Bamford. Durham, NC: Carolina Academic Press.

Barnes, J. A.

1962 African Models in the New Guinea Highlands. *Man* 62 (2): 5–9.

Battaglia, D.

1992 The Body in the Gift: Memory and Forgetting in Sabarl Mortuary Exchange. *American Ethnologist* 19 (1): 3–18.

1995 Fear of Selfing in the American Cultural Imaginary; or, "You Are Never Alone with a Clone." *American Anthropologist,* n.s. 97 (4): 672–78.

2001 Multiplicities: An Anthropologist's Thoughts on Replicants and Clones in Popular Film. *Critical Inquiry* 27: 493–514.

Beardsley, T.

1997 Cloning Hits the Big Time. *Scientific American,* Sept. 2.

Becker, A.

2003 Skepticism of Genetically Modified Crops Remains. Knight-Ridder Tribune News Service, Sept. 12.

Begley, S., et al.

1986   The Troubling Question of Fetal Rights. *Newsweek*, Dec. 8, 87.

Bird-David, N.

1993   Tribal Metaphorization of Human-Nature Relatedness: A Comparative Analysis. In *Environmentalism: The View from Anthropology*, ed. K. Milton. London: Routledge.

Birkett, D.

2002   Why Has the Response to the Black IVF Twins Born to White Parents Been So Hysterical? *Guardian*, July 10.

Blackwood, B.

1939a   Life on the Upper Watut, New Guinea. *Geographical Journal* 94 (1): 11–28.

1939b   Folk Stories of a Stone-Age People in New Guinea. *Folklore* 1 (3): 209–42.

1940   Use of Plants among the Kukukuku of South-East Central New Guinea. *Proceedings of the Sixth Pacific Science Conference* 4: 111–26. Berkeley and Los Angeles: University of California Press.

1950   *The Technology of a Modern Stone-Age People in New Guinea.* Oxford: Oxford University Press.

1978   *The Kukukuku of the Upper Watut.* Oxford: Pitt Rivers Museum.

Bledsoe, C.

2002   *Contingent Lives: Fertility, Time, and Aging in West Africa.* Chicago: University of Chicago Press.

Bloch, M.

1982   Death, Women and Power. In *Death and the Regeneration of Life*, ed. M. Bloch and J. Parry, 211–32. Cambridge: Cambridge University Press.

1992   What Goes without Saying: The Conceptualization of Zafimaniry Society. In *Conceptualizing Society*, ed. A. Kuper, 127–46. London: Routledge.

Bloch, M., and J. Parry

1982   *Death and the Regeneration of Life.* Cambridge: Cambridge University Press.

Bonavoglia, A.

1987   The Ordeal of Pamela Rae Stewart. *Ms.,* July–Aug., 92–95, 196–204.

Bonnemère, P.

1993   Maternal Nurturing Substance and Paternal Spirit: The Making of a Southern Anga Sociality. *Oceania* 64 (2): 159–86.

1996   *Le Pandanus rouge: Corps, différence des sexes et parenté chez les Ankave-Anga (Papouasie-Nouvelle-Guinée).* Paris: CNRS, Éditions de la Maison des Sciences de l'Homme.

Bordo, S.

  1993   *Unbearable Weight: Feminism, Western Culture, and the Body.*
           Berkeley and Los Angeles: University of California Press.

Bouquet, M.

  1993   *Reclaiming English Kinship: Portuguese Refractions of British Kin-
           ship Theory.* Manchester: Manchester University Press.

Bremmer, M.

  1998   Why Alien Genes Can Run Amok. *Financial Times,* Mar. 28.

Brindley, M.

  2003   Concerned Consumers Demand GM Food Labels. *Western Mail*
           (Cardiff, UK), Aug. 7.

Brock, D.

  2002   Human Cloning and Our Sense of Self. *Science* 296: 314–16.

Brosius, P.

  1997   Endangered Forest, Endangered People: Environmentalist Rep-
           resentations of Indigenous Knowledge. *Human Ecology* 25 (1):
           47–69.

  1999a   Green Dots, Pink Hearts: Displacing Politics from the Malaysian
           Rainforest. *American Anthropologist,* n.s., 101 (1): 36–58.

  1999b   Analyses and Interventions: Anthropological Engagements with
           Environmentalism. *Current Anthropology* 40 (3): 277–309.

Brown, C.

  2002   Government Backs Diane Blood's Legal Campaign. *Sunday Tele-
           graph,* Dec. 8.

Bulmer, R.

  1967   Why Is the Cassowary Not a Bird? A Problem of Zoological Tax-
           onomy among the Karam of the New Guinea Highlands. *Man,*
           n.s., 2 (1): 5–25.

Burke, J., and P. Harris

  2000   Infertile Men Turn to Fathers for Sperm. *Observer,* Nov. 19.

Burridge, K.

  1959   Siblings in Tangu. *Oceania* 30 (2): 128–54.

Burton, R., and J. Whiting

  1961   The Absent Father and Cross-Sex Identity. *Merrill-Palmer Quar-
           terly of Behavior and Development* 7 (2): 85–95.

Buttimer, A.

  1976   Grasping the Dynamism of Lifeworld. *Annals of the Association of
           American Geographers* 66 (2): 277–92.

Campbell, K., J. McWhir, W. Ritchie, and I. Wilmut

  1996   Sheep Cloned by Nuclear Transfer from a Cultured Cell Line.
           Nature 380 (6569): 64–66.

Cannell, F.

1990 Concepts of Parenthood: The Warnock Report, the Gillick Debate, and Modern Myths. *American Ethnologist* 17 (4): 667–86.

Carlin, P., and N. Biddie

1997 That Ewe, Dolly? *People,* Mar. 13, 113.

Carlson, E.

2001 Genetically Altered Organisms. *Dissent* 48 (1): 56–61.

Carsten, J.

1995 The Substance of Kinship and the Heat of the Hearth: Feeding, Personhood, and Relatedness among Malays in Pulau-Langkawi. *American Ethnologist* 22 (2): 223–41.

1997 *The Heat of the Hearth: The Process of Kinship in a Malay Fishing Community.* Oxford: Clarendon Press.

2000 Introduction: Cultures of Relatedness. In *Cultures of Relatedness: New Approaches to the Study of Kinship,* ed. J. Carsten. London: Cambridge University Press.

2001 Substantivism, Antisubstantivism, and Anti-Antisubstantivism. In *Relative Values: Reconfiguring Kinship Studies,* ed. S. Franklin and S. McKinnon. Durham, NC: Duke University Press.

2004 *After Kinship.* Cambridge: Cambridge University Press.

Cassens, D.

1987a Fetal Abuse Isn't a Crime. *American Bar Association Journal* 73 (4): 37.

1987b Is Ignoring M.D. Criminal? *American Bar Association Journal* 73 (1): 23.

Cavalli-Sforza, L., A. Wilson, C. Cantor, R. Cook-Deegan, and M. King

1991 Call for a Worldwide Survey of Human Genetic Diversity: A Vanishing Opportunity for the Human Genome Project. *Genomics* 11: 490–91.

Cavanah, S.

2002 Nature Defeats GMOs. *Mother Earth News,* Dec. 2001–Jan. 2002, 17.

Chodorow, N.

1974 Family Structure and Feminine Personality. In *Woman, Culture, and Society,* ed. M. Rosaldo and L. Lamphere, 43–66. Stanford, CA: Stanford University Press.

Cibelli, J., R. Lanza, M. West, and C. Ezzell

2002 The First Human Cloned Embryo. *Scientific American,* Jan.

Cohen, P., and M. Day

2000 Baby Born 5 Years after Father's Death. *Ottawa Citizen* (Ottawa, ON), Mar. 21.

Conservation International

1997a   www.conservation.org (accessed fall 1997)

1997b   Fact Sheet. Washington, DC: Conservation International.

1997c   Working Together to Save Our Planet. Washington, DC: Conservation International.

1998   *A Biological Assessment of the Lakekamu Basin, Papua New Guinea.* Rapid Assessment Program Working Papers 9. Washington, DC: Conservation International.

Counts, D., and D. Counts

1983   Father's Water Equals Mother's Milk: The Conception of Parentage in Kaliai West New Guinea. *Mankind* 14 (1): 46–56.

Cronon, W., ed.

1995   *Uncommon Ground: Toward Reinventing Nature.* New York: W. W. Norton.

Crumley, C., ed.

1994   *Historical Ecology: Cultural Knowledge and Changing Landscapes.* Santa Fe, NM: School of American Research Press. Distributed by the University of Washington Press.

Cussins, C.

1998   Quit Sniveling Cryo-Baby: We'll Work Out Which One's Your Mama. In *Cyborg Babies: From Techno-Sex to Techno-Tots,* ed. R. Davis-Floyd and J. Dumit. New York: Routledge.

Daley, S.

1999   The New York Times. *Pittsburgh Post,* Oct. 17.

Dalton, A.

2002   Anguish for "Father" of IVF Mix-up Baby. *Scotsman* (Edinburgh, UK), Nov. 5.

Damas, D.

1969   Contributions to Anthropology: Ecological Essays. *Proceedings of the Conference on Cultural Ecology, Ottawa, August 3–6, 1966.* Ottawa: National Museums of Canada.

Damon, F. H.

1983   Muyuw Kinship and the Metamorphosis of Gender Labor. *Man,* n.s., 18 (2): 305–26.

Darwin, C.

1979   *The Origin of Species.* New York: Gramercy Books.

Davies, K.

2001   What Makes Genetically Modified Organisms So Distasteful? *Trends in Biotechnology* 19 (10): 424–27.

Davies, M., and N. Naffine, eds.

2001   *Are Persons Property? Legal Debates about Property and Personality.* Aldershot, UK: Dartmouth Publishing.

Davis-Floyd, R., and J. Dumit, eds.

1998   *Cyborg Babies: From Techno-Sex to Techno-Tots.* New York: Routledge.

de Gruchy, S.

2002   Life, Livelihoods, and God: Why Genetically Modified Organisms Oppose Caring for Life. *Ecumenical Review* 54 (3): 251–61.

Dearle, M.

2000   Law: One Egg, Two Babies—but How Many Parents? *Independent,* Jan. 4.

Descola, P.

1992   Societies of Nature and the Nature of Society. In *Conceptualizing Society,* ed. A. Kuper, 107–26. London: Routledge.

Descola, P., and G. Pálsson, eds.

1996   *Nature and Society: Anthropological Perspectives.* London: Routledge.

Desmarais, A.

2002   Out Standing in Their Fields: Farmers Worldwide Collectively Take a Stand against Genetically Modified Organisms and Repression. *Briar Patch* 31 (5): 21–22.

DiBerardino, M., and R. McKinnell

1997   Backward Compatible. *Sciences,* Sept.–Oct., 32–37.

Doolittle, W.

1999   Phylogenetic Classification and the Universal Tree. *Science* 284: 2124–28.

Douglas, M.

1966   *Purity and Danger: An Analysis of Concepts of Pollution and Taboo.* London: Routledge and Kegan Paul.

1970   *Natural Symbols: Explorations in Cosmology.* New York: Pantheon Books.

Dunn, K.

2002   Cloning Trevor. *Atlantic Monthly* 289 (6): 31–52.

Durkheim, E., and M. Mauss

1963   *Primitive Classification.* Chicago: University of Chicago Press.

Dyer, O.

2002   Black Twins Are Born to White Parents after Treatment Mix-up. *British Medical Journal* 325 (64): 64.

Edwards, J.

2000   *Born and Bred: Idioms of Kinship and New Reproductive Technologies in England.* Oxford: Oxford University Press.

Forthcoming Skipping a Generation: Genealogy and Assisted Conception. In *Genealogy beyond Kinship: Sequence, Transmission, and Essence* in *Ethnography and Social Theory,* ed. S. Bamford and J. Leach. London: Berghahn Books.

Egan, M.

2003   A French Farmer against Big Business. *International Herald Tribune,* July 11.

Eisner, T., and E. Beiring

1994   Biotic Exploration Fund: Protecting Biodiversity through Chemical Prospecting. *BioScience* 44 (2): 95–98.

Ellen, R.

1986   What Black Elk Left Unsaid: On the Illusory Images of Green Primitivism. *Anthropology Today* 2 (6): 8–12.

Escobar, A.

1995   *Encountering Development: The Making and Unmaking of the Third World.* Princeton, NJ: Princeton University Press.

1996   Constructing Nature: Elements for a Post-Structural Political Ecology. In *Liberation Ecologies: Environment, Development, Social Movements,* ed. R. Peet and M. Watts, 46–68. London: New York: Routledge.

1998a   Environmentalism and Cultural Theory: Exploring the Role of Anthropology in Environmental Discourse. *Current Anthropology* 39 (3): 385–88.

1998b   Whose Knowledge, Whose Nature? Biodiversity Conservation and the Political Ecology of Social Movements. *Journal of Political Ecology* 5: 53–82.

1999   Steps to an Anti-Essentialist Political Ecology. *Current Anthropology* 40 (1): 1–30.

Falkner, R.

1999   Frankenstein or Benign? *World Trade Today* 55 (7): 24–26.

Feil, D. K.

1984   *Ways of Exchange: The Enga Tee of Papua New Guinea.* St. Lucia, Queensland, Australia: University of Queensland Press.

Ferguson, J.

1994   *The Anti-Politics Machine: "Development," Depoliticization, and Bureaucratic Power in Lesotho.* Minneapolis: University of Minnesota Press.

Filer, C., and W. Iamo

1989   *Baseline Planning Study for the Lake Kamu Geid Project, Gulf Province.* Mimec, Department of Anthropology and Sociology, University of Papua New Guinea. Port Moresby: Janvan.

Firth, M.

2002   IVF Twins: Black Man Is the Natural Father, White Woman Is the Natural Mother. *Evening Standard* (London), July 31.

Fischer, H.
    1968   *Negwa: Eine Papua-Gruppe im Wandel.* Mèunchen: K. Renner.
Ford, P.
    2001   Pate, Bonhomie, and a Slap at Engineered Food. *Christian Science Monitor,* Aug. 31, 1.
Fortes, M.
    1953   The Structure of Unilineal Descent Groups. *American Anthropologist* 55 (1): 17–41.
    1966   Totem and Taboo. *Proceedings of the Royal Anthropological Institute of Great Britain and Ireland* 1966: 5–22.
Foster, R.
    1990   Nature and Force-Feeding: Mortuary Feasting and the Construction of Collective Individuals in a New Ireland Society. *American Ethnologist* 17 (3): 431–48.
    1995   *Social Reproduction and History in Melanesia: Mortuary Ritual, Gift Exchange, and Custom in the Tanga Islands.* Cambridge: Cambridge University Press.
Franklin, S.
    1997   *Embodied Progress: A Cultural Account of Assisted Conception.* London: Routledge.
    1998a   Making Miracles: Scientific Progress and the Facts of Life. In *Reproducing Reproduction: Kinship, Power, and Technological Innovation,* ed. S Franklin and H. Ragoné. Philadelphia: University of Pennsylvania Press.
    1998b   The Embryo Research Debate: Science and the Politics of Reproduction. *Public Understanding of Science* 7 (3): 255–56.
    1999   The Ethics of Human Cloning. *Society* 36 (5): 103–4.
    2001a   Biological Propriety. Paper presented at the Forms of Intellectual Creativity panel at the Property, Transactions and Creations Conference, Cambridge University, Dec. 13–15.
    2001b   Biologization Revisited: Kinship Theory in the Context of the New Biologies. In *Relative Values: Reconfiguring Kinship Studies,* ed. S. Franklin and S. McKinnon, 302–25. Durham, NC: Duke University Press.
    2002   Dolly's Body: Gender, Genetics, and the New Genetic Capital. *Filozofski Vestnik* 23 (2): 119–36.
    2005   Stem Cells R Us: Emergent Life Forms and the Global Biological. In *Global Assemblages: Technology, Politics, and Ethics as Anthropological Problems,* ed. A. Ong and S. J. Collier, 59–78. Malden, MA: Blackwell Publishers.

Franklin, S., E. Hirsch, F. Price, M. Strathern, et al.

1993 *Technologies of Procreation: Kinship in the Age of Assisted Conception.* London: Routledge.

Franklin, S., C. Lury, and J. Stacey

2000 *Global Nature, Global Culture.* London: Sage.

Franklin, S., and H. Ragoné, eds.

1998 *Reproducing Reproduction: Kinship, Power, and Technological Innovation.* Philadelphia: University of Pennsylvania Press.

Frazer, J.

1911 *The Golden Bough: A Study in Magic and Religion.* London: Macmillan.

Freeman, A.

2002 White Mother's Black Twins Spark Fertility-Clinic Furor. *Globe and Mail* (Toronto), July 9.

Gallagher, J.

1987 Prenatal Invasions and Interventions: What's Wrong with Fetal Rights? *Harvard Women's Law Journal* 10: 9–58.

Garner, D.

2001 Blurring the Lines. *New York Times Magazine,* Mar. 25, 19–20.

Gillis, A.

1994 Getting a Picture of Human Diversity. *BioScience* 44 (1): 8–11.

Ginsburg, F., and R. Rapp

1991 The Politics of Reproduction. *Annual Review of Anthropology* 20: 311–43.

Godelier, M.

1982 Social Hierarchies among the Baruya of New Guinea. In *Inequality in New Guinea Highlands Societies,* ed. A. Strathern. Cambridge: Cambridge University Press.

1986 *The Making of Great Men: Male Power and Domination among the New Guinea Baruya.* Cambridge: Cambridge University Press.

Godoy, J.

2003 Rights—France: Arrest of Anti-GM Farmer Bove Sparks Protests. *Global Information Network,* June 23.

Goodale, Jane

1980 Gender, Nature, Sexuality, and Marriage: A Kaulong Model of Nature and Culture. In *Nature, Culture, and Gender,* ed. C. MacCormack and M. Strathern. Cambridge: Cambridge University Press.

Goodman, E.

1990 Pregnant and Prosecuted. *Finger Lakes Times* (Geneva, NY), Feb. 9.

Goodyear-Smith, F.
2001 Health and Safety Issues Pertaining to Genetically Modified Foods. *Australian and New Zealand Journal of Public Health* 25 (4): 371–74.

Graham, R.
2001 French Farm Leader Goes on Trial, Anti-Globalisation Movement Alleged Damage to Public Property. *Financial Times,* Feb. 8.

Guyan, C.
1999 Growing Pains. *Evening Post* (Wellington, New Zealand), May 1.

Hall, S.
2002 President's Bioethics Council Delivers. *Science* 297: 322–24.

Hanley, C.
1996 Taking a Patent on Life. *Globe and Mail* (Toronto), May 11.

Haraway, D. J.
1991 *Simians, Cyborgs, and Women: The Reinvention of Nature.* New York: Routledge.

Hardesty, D.
1977 *Ecological Anthropology.* New York: Wiley.

Hartouni, V.
1997 *Cultural Conceptions: On Reproductive Technologies and the Remaking of Life.* Minneapolis: University of Minnesota Press.

Hayden, C.
1998 A Biodiversity Sampler for the Millennium. In *Reproducing Reproduction: Kinship, Power, and Technological Innovation,* ed. S. Franklin and H. Ragoné, 173–206. Philadelphia: University of Pennsylvania Press.

Henig, R.
2003a After 25 Years, Birth Issues Linger. *Newsday,* July 16.
2003b Pandora's Baby. *Scientific American,* June, 62–67.
2004 *Pandora's Baby: How the First Test-Tube Babies Sparked the Reproductive Revolution.* Boston: Houghton Mifflin.

Henley, J.
2003 Life: Will the President Pardon Bove? *Guardian,* June 26.

Hennelly, J.
1912 *Papuan Annual Report.* National Archives, Boroko, Papua New Guinea.

Herdt, G.
1981 *Guardians of the Flutes: Idioms of Masculinity.* New York: McGraw-Hill.
1987 *Sambia: Ritual and Gender in New Guinea.* New York: Holt, Rinehart and Winston.

1996   Ritual Rebirth. Paper presented at the Conference on the Angan Peoples of Papua New Guinea. Aix-en-Provence, France.

Herscovici, A.

1985   *Second Nature: The Animal Rights Controversy.* Montreal: CBC Enterprises.

Hertz, R.

1960   *Death and the Right Hand.* Glencoe, IL: Free Press.

Higginson, C.

1908   *Papua Annual Report.* National Archives, Boroko, Papua New Guinea.

Hiscock, J.

1997   Surrogate Mom Having Dead Woman's Baby. *Standard* (St. Catherines, ON), Dec. 9.

Holden, C., and J. Kaiser

2002   Report Backs Ban: Ethics Panel Debuts. *Science* 295: 601–2.

Holy, L.

1996   *Anthropological Perspectives on Kinship.* London: Pluto Press.

Home, C.

1999   A Baby That Brings Horror Story to Life. *Daily Mail* (London). Mar. 26.

Huntington, R., and P. Metcalf

1979   *Celebrations of Death: The Anthropology of Mortuary Ritual.* Cambridge: Cambridge University Press.

Ingold, T.

1990   An Anthropologist Looks at Biology. *Man,* n.s., 25 (2): 208–29.

1991   Becoming Persons: Consciousness and Sociality in Human Evolution. *Cultural Dynamics* 4: 355–78.

1992   Culture and the Perception of the Environment. In *Bush, Base, Forest, Farm: Culture, Environment, and Development,* ed. E. Croll and D. Parkin. London: Routledge.

1993   The Temporality of the Landscape. *World Archaeology* 25 (2): 152–74.

1994   From Trust to Domination: An Alternative History of Human-Animal Relations. In *Animals and Human Society: Changing Perspectives,* ed. A. Manning, J. Serpell, and R. Edinburgh, 1–22. London: Routledge.

2000   *The Perception of the Environment: Essays in Livelihood, Dwelling, and Skill.* London: Routledge.

Ivins, M.

1999   A Genetic Engineering Controversy. *Buffalo Times* (Buffalo, NY), Jan. 6.

Jaenisch, R., and I. Wilmut

    2001   Don't Clone Humans! *Science,* Mar. 30, 291.

James, S., and S. Palmer, eds.

    2002   *Visible Women: Essays on Feminist Legal Theory and Political Philosophy.* Portland, OR: Hart Publishing.

Jamieson, R.

    1999   French Farmer Bove Takes His Beef to McDonalds. *Seattle Post,* Nov. 30.

Jorgensen, Dan.

    1981   Taro and Arrows: Order, Entropy, and Religion among the Telefolmin. Ph.D. diss., University of British Columbia.

Joyce, C.

    1992   Western Medicine Men Return to the Field. *BioScience* 42 (6): 399–403.

Kaberry, P.

    1967   The Plasticity of New Guinea Kinship. In *Social Organization: Essays Presented to Raymond Firth,* ed. M. Freedman. Chicago: University of Chicago Press.

Kahn, M.

    1990   Stone-Faced Ancestors: The Spatial Anchoring of Myth in Wamira, Papua New Guinea. *Ethnology* 29 (1): 51–66.

Kahn, S.

    2000   *Reproducing Jews: A Cultural Account of Assisted Conception in Israel.* Durham, NC: Duke University Press.

Kaiser, J.

    2000   Panel Urges Further Study of Biotech Corn. *Science,* Dec. 8, 1867.

Kass, L.

    1971   The New Biology: What Price Relieving Man's Estate? *Science* 174: 779–88.

    1972   Making Babies: The New Biology and the Old Mortality. *Public Interest* 26 (Winter): 18–56.

    1989   "Making Babies" Revisited. In *Ethical Issues in Modern Medicine,* 3rd ed., ed. J. Arros and N. Rhoden. Mountain View, CA: Mayfield Publishing.

    1997   The Wisdom of Repugnance. *New Republic* 216 (22): 17–26.

Kelly, R.

    1977   *Etoro Social Structure: A Study in Structural Contradiction.* Ann Arbor: University of Michigan Press.

Kenward, M.

    1994   Pure Geneius. *Director* 48 (3): 54–58.

Kirsch, S.

    1997   Regional Dynamics and Conservation in Papua New Guinea:

The Lakekamu-Kunimaipa Basin Project. *Contemporary Pacific* 9 (1): 97–120.

Klee, K.

1999   Europeans Are Rallying against "Frankenstein" Foods. *Newsweek,* Sept. 13, 22.

Knestout, B.

2000   Food Fight. *Kiplinger's Personal Finance Magazine* 54 (Jan.): 56.

Kohl, D.

2001   GM Food: Another View. *The Nation,* Apr. 16, 7, 23.

Kolder, V., J. Gallagher, and M. Parsons

1987   Court-Ordered Obstetrical Interventions. *New England Journal of Medicine* 316 (19): 1192–96.

Lambek, M.

1992   Taboo as Cultural-Practice among Malagasy Speakers. *Man,* n.s., 27 (2): 245–66.

Lambert, B.

1983   Authority, Equality, and Complementarity in Butaritari-Makin Sibling Relationships (Northern Gilbert Islands). In *Siblingship in Oceania: Studies in the Meaning of Kin Relations,* ed. M. Marshall. ASAO monograph no. 8. Ann Arbor: University of Michigan Press.

Langness, L.

1967   Sexual Antagonism in New-Guinea Highlands: Bena Bena Example. *Oceania* 37 (3): 161–77.

Latour, B.

1993   *We Have Never Been Modern.* Cambridge, MA: Harvard University Press.

Leach, E.

1964   Anthropological Aspects of Language: Animal Categories and Verbal Abuse. In *New Directions in the Study of Language,* ed. E. H. Leeneberg. Cambridge: MIT Press.

Leach, J.

2003   *Creative Land: Place and Procreation on the Rai Coast of Papua New Guinea.* Oxford: Berghahn Books.

Leary, W.

2002   Panel Urges Caution in Producing Gene-Altered Animals. *New York Times,* Aug. 21.

Lee, R., and I. DeVore

1968   *Man the Hunter.* Chicago: Aldine Publishing.

Leenhardt, M.

1979   *Do Kamo: Person and Myth in the Melanesian World.* Trans. B. M. Gulati. Chicago: University of Chicago Press.

Lemonnier, P.
   1991   From Great Men to Big Men: Peace, Substitution, and Compe-
               tition in the Highlands of New Guinea. In *Big Men and Great
               Men: Personifications of Power in Melanesia,* ed. M. Godelier and
               M. Strathern. Cambridge: Cambridge University Press.

Lévi-Strauss, C.
   1961   Tristes Tropiques: From an Anthropologists Memoirs. *Encounter*
               16 (2): 7–23.
   1966   *The Savage Mind.* Trans. J. Weightman and D. Weightman. Chi-
               cago: University of Chicago Press.
   1967   *Totemism.* Trans. R. Needham. Boston: Beacon Press.
   1969   *The Elementary Structures of Kinship.* Trans. J. Bell and J. von
               Sturmer. Boston: Beacon Press.
   1976   *Structural Anthropology.* Trans. M. Layton. Vol. 2. Chicago: Uni-
               versity of Chicago Press.

Lewin, R.
   1993   Genes from a Disappearing World. *New Scientist* 29 (1875): 25–30.

Lewin, T.
   1987   Courts Acting to Force Care on the Unborn. *New York Times.*
               Nov. 23.

Lewontin, R.
   2000   *It Ain't Necessarily So: The Dream of the Human Genome and
               Other Illusions.* New York: New York Review Books.

Lilliston, B.
   2001   Farmers Fight to Save Organic Crops. *The Progressive,* Aug.,
               26–29.

LiPuma, E.
   1988   *The Gift of Kinship: Structure and Practice in Maring Social Orga-
               nization.* Cambridge: Cambridge University Press.

Lofstedt, R.
   2003   The Ideas Exchange: Expert View: Plant New Seeds in the GM
               Debate. *Independent on Sunday* (London), Oct. 5.

Luke, T.
   1997   The World Wildlife Fund: Eco-Colonialism as Funding the
               World Wide "Wise-Use" of Nature. *Capitalism, Nature, Social-
               ism: A Journal of Socialist Ecology* 8 (2): 31–61.

Lutz, C., and J. Collins
   1993   *Reading "National Geographic."* Chicago: University of Chicago
               Press.

Lyall, S.
   2002   Whites Have Black Twins in In-Vitro Mix-up. *New York Times,*
               July 9.

MacCormack, C., and M. Strathern, eds.

1980   *Nature, Culture, and Gender.* Cambridge: Cambridge University Press.

Maeder, J.

2001   Chemistry Set: The Baby That Wasn't. *New York Daily News,* Oct. 7.

Maine, H.

1861   *Ancient Law.* London: John Murray.

Mallet, V.

2001   French Protestors Destroy Modified Maize. *Financial Times,* Aug. 27.

Marks, J.

1995   The Human Genome Diversity Project: Good for if Not Good as Anthropology. *Anthropology Newsletter,* Apr.

Marsh, B.

2002   Three-Way Mix-up Gives IVF Patients the Wrong Embryos. *Daily Mail* (London), Oct. 28.

Marshall, M., ed.

1981a   Approaches to Siblingship in Oceania. Introduction to *Siblingship in Oceania: Studies in the Meaning of Kin Relations,* ed. M. Marshall, 1–15. Ann Arbor: University of Michigan Press.

1981b   *Siblingship in Oceania: Studies in the Meaning of Kin Relations.* Lanham, MD: University Press of America.

Marwick, C.

1997   Scientists Flock to Hear Cloner Wilmut at the NIH. *Journal of the American Medical Association* 277 (14): 1102.

Marx, K.

1976   *Capital.* Trans. B. Fowkes. Vol. 1. New York: Vintage Books.

Marx, K., and F. Engles

1970   *The German Ideology.* Ed. C. J. Arthur. New York: International Publishers.

Maschio, T.

1994   *To Remember the Faces of the Dead: The Plenitude of Memory in Southwestern New Britain.* Madison: University of Wisconsin Press.

Mayr, E.

1976   *Evolution and the Diversity of Life.* Cambridge, MA: Harvard University Press.

1982   *The Growth of Biological Thought: Diversity, Evolution, and Inheritance.* Cambridge, MA: Belknap Press.

Mbaginta, O.

1976   Medical Practices and Funeral Ceremony of the Dunkwi Anga. *Journal de la Société des Océanistes* 32 (53): 299–305.

McCarthy, J.
    1964   *Patrol into Yesterday: My New Guinea Years.* Melbourne: F. W. Cheshire.

McEnroe, C.
    2000   A Page from History We Were There: July 26, 1978 Series. *Hartford Courant,* Mar. 26.

McHughen, A.
    2000   *Pandora's Picnic Basket: The Potential and Hazard's of Genetically Modified Foods.* Oxford: Oxford University Press.

McKillop, J.
    2002   Second IVF Mix-up Couple Still Childless: Anguish for Father of Black Twins Born to White Parents. *Herald* (Glasgow, UK), Nov. 5.

McKinnon, S., and S. Silverman, eds.
    2005   *Complexities: Beyond Nature and Nurture.* Chicago: University of Chicago Press.

McLean, M.
    2002   Seeing Double: The Ethics of Human Cloning. *Hasting's Law Journal* 53 (5): 1049–56.

McNamara, E.
    1989   Fetal Endangerment Cases on the Rise. *Boston Globe,* Oct. 3.

Meggitt, M.
    1972   System and Subsystem: Te Exchange Cycle among Mae Enga. *Human Ecology* 1 (2): 111–23.

Mestel, R.
    2003   Birth by Test Tube Turns 25. *Los Angeles Times,* July 24.

Milton, K.
    1993   Environmentalism: The View from Anthropology. London: Routledge.

Mimica, J.
    1981   *Omalyce: An Ethnography of the Ikwaye View of the Cosmos.* Ph.D. diss., Australian National University, Canberra.

    1988   *Intimations of Infinity: The Mythopoeia of the Iqwaye Counting System and Number.* Oxford: Berg.

    1991   The Incest Passions: An Outline of the Logic of the Iqwaye-Social-Organization. *Oceania* 62 (2): 81–113.

Morgan, L.
    1877   *Ancient Society: Researches in the Lines of Human Progress from Savagery through Barbarism to Civilization.* New York: Henry Holt.

Mosko, M.
    1983   Conception, De-Conception, and Social-Structure in Bush Mekeo Culture. *Mankind* 14 (1): 24–32.

Mulugu, I.

1998 Are We Creating New Frankensteins? Taking Over from Where Nature Left Off. *Businessline,* June 22.

Mundy, L.

2004 Present at the Creation: The Strange Saga of the Latter-Day Baby-Making Revolution. *Washington Post,* Feb. 8.

Munn, N.

1970 The Transformation of Subjects into Objects in Walbiri and Pitjantjatjara Myth. In *Australian Aboriginal Anthropology,* ed. R. Berndt. Nedlands, Western Australia: University of Western Australian Press.

1973 *Walbiri Iconography: Graphic Representation and Cultural Symbolism in a Central Australian Society.* Chicago: University of Chicago Press.

1986 *The Fame of Gawa: A Symbolic Study of Value Transformation in a Massim (Papua New Guinea) Society.* Cambridge: Cambridge University Press.

Myers, F.

1986 Pintupi Country, Pintupi Self: Sentiment, Place, and Politics among Western Desert Aborigines. Washington, DC: Smithsonian Institution Press.

Naffine, N.

2002 Can Women Be Legal Persons? In *Visible Women: Essays on Feminist Legal Theory and Political Philosophy,* ed. S. James and S. Palmer, 69–90. Portland, OR: Hart Publishing.

Nash, M.

2000 Grains of Hope. *Time,* July 31, 14–21.

Nelkin, D., and S. Lindee

1995 *The DNA Mystique: The Gene as a Cultural Icon.* New York: W. H. Freeman.

Nelson, L., B. Buggy, and C. Weil.

1986 Forced Medical Treatment of Pregnant Women: Compelling Each to Live as Seems Good to the Rest. *Hastings Law Journal* 37 (5): 703–63.

Nichols, J.

1999 French Farmer Is Leading Man at WTO. *Madison Capital Times* (Madison, WI), Nov. 30.

Niles, T.

2001 The Illusion of "Frankenfood." *Washington Times,* Nov. 7.

Nothnagel, D.

1996 The Reproduction of Nature in Contemporary High-Energy

Physics. In *Nature and Society: Anthropological Perspectives,* ed. P. Descola and G. Pálsson. London: Routledge.

Nuttall, N.
1998   Silent Spring, 2020: Mind and Matter. *Times* (London), July 13.

Oates, W., and L. Oates
1968   *Kapau Pedagogical Grammar.* Canberra: Australian National University.

Ortner, S.
1974   Is Female to Male as Nature Is to Culture? In *Woman, Culture, and Society,* ed. M. Rosaldo and L. Lamphere. Stanford, CA: Stanford University Press.

Ortner, S., and H. Whitehead
1981   Introduction to *Sexual Meanings: The Cultural Construction of Gender and Sexuality,* ed. S. Ortner and H. Whitehead. Cambridge: Cambridge University Press.

Pálsson, G.
2005   The Web of Kin: An Online Genealogical Machine. In *Genealogy Beyond Kinship: Sequence, Transmission, and Essence in Ethnography and Social Theory,* ed. S. Bamford and J. Leach. Oxford: Berghahn Books.

Papagaroufali, E.
1996   Xenotransplantation and Transgenesis: Immoral Stories about Human-Animal Relations in the West. In *Nature and Society: Anthropological Perspectives,* ed. P. Descola and G. Pálsson. London: Routledge.

1999   Donation of Human Organs or Bodies after Death: A Cultural Phenomenology of "Flesh" in the Greek Context. *Ethos* 27 (3): 283–314.

Parkin, R.
1997   *Kinship: An Introduction to Basic Concepts.* Oxford: Blackwell Publishers.

Pence, G.
1998   Happy Birthday, Louise: First Test-Tube Baby Is Now 20 and the Sky Hasn't Fallen. *Gazette* (Montreal), July 25.

Pennisi, E.
2003   Passages Found through Labyrinth of Bacterial Evolution. *Science* 301: 745–46.

Pennisi, E., and N. Williams
1997   Will Dolly Send in the Clones? *Science* 275: 1415–16.

Pepper, A.
1996   Victim of Leukemia Leaves Embryos. *Oregonian* (Portland), Dec. 22.

Peres, J.

1997a   Surrogate Mom with Dead Woman's Baby Raises Moral Issues. *Buffalo News* (Buffalo, NY), Dec. 15.

1997b   A Mother Beyond the Grave? Parents Have Ova of Dead Daughter Planted in Surrogate. *Seattle Times*, Dec. 15.

Perlman, D.

1993   A Search among Vanishing Peoples: Genetic Sleuths Race against Time. *San Francisco Chronicle*, Apr. 21.

Pickrell, J.

2001   Experts Assail Plans to Help Childless Couples. *Science* 291: 2061–63.

Pollan, M.

2001   Genetic Pollution. *New York Times Magazine,* Dec. 9, 74, 76.

Pollitt, K.

1990   Fetal Rights: A New Assault on Feminism. *The Nation,* Mar. 26, 409.

Pook, S., and N. Martin

2002   Staff Shortage Blamed for IVF Clinic Mix-up. *Daily Telegraph* (London), Oct. 29.

Poole, F.

1981   Transforming "Natural" Woman: Female Ritual Leaders and Gender Ideology among Bimin-Kuskusmin. In *Sexual Meanings: The Cultural Construction of Gender and Sexuality,* ed. S. W. Ortner and H. Whitehead. Cambridge: Cambridge University Press.

Pottage, A.

1998   Genes, Patents, and Bio-Politics. *Modern Law Review* 61 (5): 740–65.

n.d. a   An Original Genetic Inheritance. E-mail copy of unpublished paper, provided by author.

n.d. b   Terminator. E-mail copy of unpublished paper, provided by author. A shorter version was presented at the Property, Transactions and Creations Conference, Cambridge University, Dec. 13–15.

Pouwer, J.

1960   "Loosely Structured Societies" in Netherlands New Guinea. *Bijdragen tot de Taal-, Land- en Volkenkunde* 116 (1): 109–18.

President's Council on Bioethics

2002   *Human Cloning and Human Dignity: An Ethical Inquiry.* President's Council on Bioethics. www.bioethics.gov/reports/cloning report/index.html.

Provincial Data System Community Register

1978   *Gulf Province.* National Statistics Office, Wards Strip, Papua New Guinea.

Rabinow, P.

1992   Artificiality and Enlightenment: From Sociobiology to Biosociality. In *Incorporations: Zone 6,* ed. J. Crary and S. Kwinter. New York: Urzone.

1996   *Making PCR: A Story of Biotechnology.* Chicago: University of Chicago Press.

Radcliffe-Brown, A. R.

1950   Introduction to *African Systems of Kinship and Marriage,* ed. A. R. Radcliffe-Brown and C. Forde. London: Oxford University Press.

Ragoné, H.

1994   *Surrogate Motherhood: Conception in the Heart.* Boulder, CO: Westview Press.

Rao, R.

2002   What's So Strange about Human Cloning? *Hastings Law Journal* 53 (5): 1007–16.

Rapp, R.

1999   *Testing Women, Testing the Fetus: The Social Impact of Amniocentesis in America.* New York: Routledge.

Rappaport, R.

1968   *Pigs for the Ancestors.* New Haven, CT: Yale University Press.

1979   *Ecology, Meaning, and Religion.* Berkeley, CA: North Atlantic Press.

Raven, P.

1990   The Politics of Preserving Biodiversity. *BioScience* 40 (10): 769–73.

Read, K.

1952   Nama Cult of the Central Highlands, New Guinea. *Oceania* 23 (1): 1–25.

1954   Cultures of the Central Highlands, New Guinea. *Southwestern Journal of Anthropology* 10 (1): 1–43.

Rhoden, N. K.

1986   The Judge in the Delivery Room: The Emergence of Court-Ordered Cesareans. *California Law Review* 74 (6): 1951–2030.

Ripston, R.

1990   One Baby, Three Parents: Whose Rights Prevail? *Los Angeles Times,* Oct. 17.

Rival, L.

1993   The Growth of Family Trees: Understanding Huaorani Perceptions of the Forest. *Man,* n.s., 28 (4): 635–52.

1998   Androgynous Parents and Guest Children: The Huarani Cou-
vade. *Journal of the Royal Anthropological Institute* 5 (4): 619–42.

Rivers, W. H. R.

1968   The Genealogical Method of Anthropological Inquiry. In *Kinship
and Social Organization,* ed. W. H. R. Rivers. New York: Athlone
Press.

Riviere, P.

1971   Marriage: A Reassessment. In *Rethinking Kinship and Marriage,*
ed. R. Needham. ASA Monographs no. 11. London: Tavistock
Publications.

Roberts, L.

1991   A Genetic Survey of Vanishing Peoples. *Science* 252: 1614–17.

Rodman, M.

1987   *Masters of Tradition: Consequences of Customary Land Tenure in
Longana, Vanuatu.* Vancouver: University of British Columbia
Press.

Rohatynskyj, M.

1990   The Larger Context of Omie Sex Affiliation. *Man,* n.s., 23 (3):
434–53.

Rorvik, D.

1971   The Test Tube Baby Is Coming: Taking Life in Our Own Hands.
Look, May 18, 82.

Rosaldo, M.

1974   Women, Culture, and Society: A Theoretical Overview. In
Woman, Culture, and Society, ed. M. Rosaldo and L. Lamphere.
Stanford, CA: Stanford University Press.

Rosenfeld, A., and L. Harris

1969   The Second Genesis: Science, Sex, and Tomorrow's Morality.
Live, 37–54.

Rowland, R.

1992   *Living Laboratories: Women and Reproductive Technologies.* Bloom-
ington: Indiana University Press

Rozenberg, J., and P. Stokes

2002   Couple in IVF Mix-up "Dearly Love Black Twins." *Daily Telegraph* (London), Nov. 5.

Rubin, G.

1975   The Traffic in Women: Notes on the "Political Economy" of Sex.
In *Toward an Anthropology of Women,* ed. R. Reiter, 157–210. New
York: Monthly Review Press.

Rubin, T.

1999   Food Fight! Europeans Battling Genetically Modified U.S. Exports. *Tulsa World* (Tulsa, OK), Oct. 17.

Rugg, K.

2002   Understanding the GMO Issue. *Bangkok* (Thailand) *Post,* June 9.

Said, E.

1978   Orientalism. New York: Random House.

Sancton, T.

1999   Super Fries Saboteur. *Time,* Dec. 6, 74.

Savell, K.

2002   The Mother of the Legal Person. In *Visible Women: Essays on Feminist Legal Theory and Political Philosophy,* ed. S. James and S. Palmer, 29–68. Portland, OR: Hart Publishing.

Schieffelin, E.

1976   *The Sorrow of the Lonely and the Burning of the Dancers.* New York: St. Martin's Press.

Schneider, D.

1968   *American Kinship: A Cultural Account.* Englewood Cliffs, NJ: Prentice-Hall.

1984   *A Critique of the Study of Kinship.* Ann Arbor: University of Michigan Press.

Sciclino, E.

2002   French Foe of Globalism Gets 14 Months in Jail. *Chicago Tribune,* Nov. 22.

Seamark, M.

2002   Black Twins Staying Put: White Parents Will Keep Mix-up Babies. *Daily Mail* (London), Nov. 5.

Segal, N.

2002   Human Cloning: Insights from Twins and Twin Research. *Hastings Law Journal* 53 (5): 1073–84.

Shapiro, H.

1997   Ethical and Policy Issues of Human Cloning. *Science* 277: 195–96.

Shapiro, J.

1985   The Sibling Relationship in Lowland South America: General Considerations. In *The Sibling Relationship in Lowland South America,* ed. K. Kensinger, vol. 7. Working papers on South America Indians. Bennington, VT: Bennington College.

Shaw, M.

1984   Conditional Prospective Rights of the Fetus. *CSJ J. Legal Med.* 63: 96–100.

Shiva, V.

2000   *Stolen Harvest: The Hijacking of the Global Food Supply.* Cambridge, MA: South End Press.

Shore, C.

1992   Virgin Births and Sterile Debates: Anthropology and the New Reproductive Technologies. *Current Anthropology* 33 (3): 295–314.

Siegelitzkovich, J.

1998   The Heartbreaking Legacy of Julie Garber. *Jerusalem Post,* Feb. 1.

Sillitoe, P.

1979   *Give and Take: Exchange in Wola Society.* Canberra: Australian National University Press.

Silver, L.

1997   *Remaking Eden: Cloning and Beyond in a Brave New World.* New York: Avon Books.

Simpson, C.

1953   *Adam with Arrows: Inside Aboriginal New Guinea.* New York: Praeger.

Sinclair, J.

1961   Patrolling in the Territory of Papua and New Guinea. *Australian Territories* 1 (4): 26–33.

1966   *Behind the Ranges: Patrolling in New Guinea.* Melbourne: Melbourne University Press.

Smith, L., and O. Wright

2002   White Couple in IVF Blunder Can Keep Black Twins. *Times* (London), Nov. 5.

Soules, M.

1999   Commentary: Posthumous Harvesting of Gametes; A Physician's Perspective. *Journal of Law, Medicine, Ethics* 27 (4): 362–65.

Stafford, C.

2000   Chinese Patriliny and the Cycles of Yang and Laiwang. In *Cultures of Relatedness: New Approaches to the Study of Kinship,* ed. J. Carsten. Cambridge: Cambridge University Press.

Stanworth, M.

1987   Introduction, and Reproductive Technologies and the Deconstruction of Motherhood. In *Reproductive Technologies: Gender, Motherhood, and Medicine,* ed. M. Stanworth, 1–35. Cambridge: Cambridge University Press.

Strathern, A.

1971a   Wiru and Daribi Matrilateral Payments. *Journal of the Polynesian Society* 80 (4): 449–62.

1971b   *The Rope of Moka: Big-Men and Ceremonial Exchange in Mount Hagen, New Guinea.* London: Cambridge University Press.

1973   Kinship, Descent, and Locality: Some New Guinea Examples. In *The Character of Kinship,* ed. J. Goody and M. Fortes. London: Cambridge University Press.

1981   Death as Exchange: Two Melanesian Cases. In *Morality and Immorality,* ed. H. King. London: Academic Press.

Strathern, M.

1987   Producing Difference: Connections and Disconnections in two New Guinea Highland Kinship Systems. In *Gender and Kinship: Essays Toward a Unified Analysis,* ed. J. F. Collier and S. J. Yanagisako. Stanford, CA: Stanford University Press.

1988   *The Gender of the Gift: Problems with Women and Problems with Society in Melanesia.* Berkeley and Los Angeles: University of California Press.

1992a   *After Nature: English Kinship in the Late Twentieth Century.* Cambridge: Cambridge University Press.

1992b   *Reproducing the Future: Essays on Anthropology, Kinship, and the New Reproductive Technologies.* Manchester: Manchester University Press.

1993   Making Incomplete. In *Carved Flesh/Cast Selves: Gendered Symbols and Social Practices,* ed. V. Broche-Due, I. Rudie, and T. Bleie. Oxford: Berg.

1995   Future Kinship and the Study of Culture. *Futures* 27 (4): 423–35.

1997   Parenthood in Modern Society: Legal and Social Issues for the Twenty-first Century. *Journal of the Royal Anthropological Institute* 3 (2): 409–10.

2001   Same-Sex and Cross-Sex Relations: Some Internal Comparisons. In *Gender in Amazonia and Melanesia: An Exploration of the Comparative Method,* ed. T. A. Gregor and D. Tuzin. Berkeley and Los Angeles: University of California Press.

Sykes, K.

2006   "Family Planning": The Politics of Reproduction in Central New Ireland. In *Embodying Modernity and Postmodernity: Ritual, Praxis, and Social Change in Melanesia,* ed. S. Bamford. Durham, NC: Carolina Academic Press.

Syvanen, M.

2002   Genes on the Move. *American Scientist* 90 (4): 380.

Takagi, S.

2000   Incest at Root of Sperm Donor Row. *Mainichi Daily News* (Tokyo), Sept. 19.

Tangley, L.

1990   Cataloging Costa Rica's Diversity. *BioScience* 40 (6): 633–36.

Taylor, L.

2002   Clinic's Mistake Shakes In Vitro Birth Industry. *Toronto Star,* July 10.

Taylor, M.

2002   Full Heartache of IVF Twins Error Revealed. *Daily Post* (Liverpool, UK), Nov. 5.

Toner, M.

2002   What's Coming to Dinner? Seeds of Change. *Atlanta Journal,* May 19.

Tuzin, D.

1972   Yam Symbolism in the Sepik: An Interpretive Account. *Southwestern Journal of Anthropology* 28 (3): 230–54.

Van Gennep, A.

1960   *The Rites of Passage.* Chicago: University of Chicago Press.

Vayda, A.

1969   *Environment and Cultural Behavior: Ecological Studies in Cultural Anthropology.* Garden City, NY: Natural History Press.

Verkaik, R.

2002   White Woman Is Real Mother of Black Twins. *Independent,* Aug. 1.

Viveiros de Castro, E.

Forthcoming   The Gift and the Given: Three Nano-Essays on Kinship and Magic. In *Genealogy Beyond Kinship: Sequence, Transmission, and Essence in Ethnography and Social Theory,* ed. S. Bamford and J. Leach. Oxford: Berghahn Books.

Vogel, G.

2001   Human Cloning Plans Spark Talk of U.S. Ban. *Science* 292: 31.

2002   U.N. Split over Full or Partial Cloning Ban. *Science* 298: 1316–17.

Wagner, R.

1967   *The Curse of Souw: Principles of Daribi Clan Definition and Alliance in New Guinea.* Chicago: University of Chicago Press.

1972   *Habu: The Innovation of Meaning in Daribi Religion.* Chicago: University of Chicago Press.

1974   Are There Social Groups in the New Guinea Highlands? In *Frontiers of Anthropology: An Introduction to Anthropological Thinking,* ed. M. J. Leaf and B. G. Campbell. New York: Van Nostrand.

1975   *The Invention of Culture.* Englewood Cliffs, NJ: Prentice-Hall.

1977   Analogic Kinship: A Daribi Example. *American Ethnologist* 4 (4): 623–42.

n.d.   Misreading the Metaphor: Cross-Cousin Relationships in the New Guinea Highlands. Unpublished paper provided by author.

Wallace, B.

1997   The Dolly Debate. *Maclean's* (Toronto), Mar. 10, 54.

Wallis, C.

1984   Making Babies: The New Science of Conception. *Time,* Sept. 10, 46–56.

Ward, A.

2002   Life after Death: Like Diane Blood, These Woman Had Their Children after Their Husbands Died. So How Do They Feel Bringing Up a Living Reminder of What They've Lost? *Daily Mail* (London), Apr. 2.

Watson, J.

1965   Loose Structure Loosely Construed: Groupless Groupings in Gadsup? *Oceania* 35: 267–71.

Weinberg, S.

2004   Reproductive Technology: Today and Tomorrow. *Milwaukee Journal Sentinel,* Jan. 25.

Weiner, A.

1992   *Inalienable Possessions: The Paradox of Keeping-While-Giving.* Berkeley and Los Angeles: University of California Press.

1980   Reproduction: A Replacement for Reciprocity. *American Ethnologist* 7 (1): 71–85.

Weiner, J.

1982   Substance, Siblingship, and Exchange: Aspects of Social Structure in New Guinea. *Social Analysis* 11 (Oct.): 3–34.

Weismantel, M.

1995a   Making Kin: Kinship Theory and Zumbagua Adoptions. *American Ethnologist* 22 (4): 685–704.

1995b   Response to McKinnon. *American Ethnologist* 22 (4): 706–9.

Weiss, R.

1998a   Babies in Limbo: Laws Outpaced by Fertility Advances. *Washington Post,* Dec.

1998b   Women's Health: The Many Bumps on the Road to Babyville: Modern Technology and Lack of Regulation Have Combined to Create a U.S. Fertility Industry Rife with Moral Dilemmas. *Los Angeles Times,* Mar. 23.

2005   The Power to Divide. *National Geographic,* July, 3–27.

Weston, K.

1995   Forever Is a Long Time: Romancing the Real in Gay Kinship Ideologies. In *Naturalizing Power: Essays in Feminist Cultural Analysis,* ed. S. J. Yanagisako and C. Delaney. New York: Routledge.

2001   Kinship, Controversy, and the Sharing of Substance: The Race/Class Politics of Blood Transfusion. In *Relative Values: Re-*

configuring Kinship Studies, ed. S. Franklin and S. McKinnon. Durham, NC: Duke University Press.

Wills, C.
1998   A Sheep in Sheep's Clothing? *Discover,* Jan., 22–23.

Wilmut, I., K. Campbell, and C. Tudge
2000   *The Second Creation: Dolly and the Age of Biological Control.* New York: Farrar, Straus and Giroux.

Wilmut, I., B. De Sousa, A. Dinnyes, et al.
2002   Somatic Cell Nuclear Transfer. *Nature* 419 (6709): 583–86.

Wilmut, I., A. Schmicke, E. McWhir, J. Kind, and K. Campbell
1997   Viable Offspring Derived from Fetal and Adult Mammalian Cells. *Nature* 385 (6619): 810–13.

Wilson, D.
1984   Saving the Embryos. *Boston Globe,* June 26.

Winston, R.
2002   Comment and Analysis: Blaming IVF Doctors Alone Is Wrong; The Birth of Black Twins to White Parents Is an Error Not a Tragedy. *Guardian,* July 10.

Wolfram, S.
1987   *In-laws and Outlaws: Kinship and Marriage in England.* London: Croom Helm.

Wolinsky, H.
1988   In Vitro Fertilization Still Stirs Questions. *Chicago Sun,* July 10.

Wulf, J.
1999   What If My Test-Tube Babies Were Swapped in the Lab? *Time,* Apr. 12, 69.

Yanagisako, S., and J. Collier
1987   Toward a Unified Analysis of Gender and Kinship. In *Gender and Kinship: Essays Toward a Unified Analysis,* ed. J. F. Collier and S. J. Yanagisako. Stanford, CA: Stanford University Press.

Zerner, C.
1994   Telling Stories about Biological Diversity. In *Valuing Local Knowledge: Indigenous People and Intellectual Property Rights,* ed. S. B. D. S. Brush, 68–101. Washington DC: Island Press.

2000   *People, Plants, and Justice: The Politics of Nature Conservation.* New York: Columbia University Press.

Zimmer, G.
1969   When the Kukukuku Came from the Hills It Was to Kill. *Pacific Islands Monthly* 40 (11): 85–93.

INDEX

abortion rights, 89–90

ACTs (assisted conception technologies). *See* reproductive technologies

adoption, 47, 57

AEBC (Agriculture and Environment Biotechnology Commission), 23–24

affinity. *See* marriage

Agriculture and Environment Biotechnology Commission (AEBC), 23–24

Akeanga myth, 36–39, 109, 137–38

Alltown (England), 53, 183n13

am'a pit'ya apa (memorial songs), 128–29

am'a pit'ya itya iyamakio. See mortuary feasts

American Kinship: A Cultural Account (Schneider), 7–8

American Society of Reproductive Medicine, 51

American Type Culture Collection (Rockville, Maryland), 163, 168

anaka (pandanus nut trees), 33–34

Angan people, 16–17, 114, 180n10. *See also* Kamea men; Kamea women

Anheuser-Busch, 157

Ankave people, 180n10

anthropological studies: cross-cultural focus of, 175–76; juxtaposed to Kamea model, 173–74; of kinship formulations, 7–10, 56–58, 175–76; of matrilateral payments, 64–65, 184nn17,18; of mortuary symbolism, 124–27, 138, 144, 148–49; mortuary symbolism theories in, 11; of siblingship, 59–60, 183n13; socialized nature theory in, 28–29, 181–82n8, 182n9; of taboos, 91, 92

apa. See male initiation

apaka. See Kamea women

"Are Mothers Persons?" (Bordo), 84

arm bands *(ituka)*, 107

assisted conception technologies (ACTs). *See* reproductive technologies

Assisted Conception Unit of Leeds General Infirmary, 46

Associated Press, 154

Auch (France), 20

awati apaka ("good woman"), 33

Bacillus thuringiensis (Bt), 21, 181nn3,6

Balisby, Sam, 83

cane (*pitpit*), 32, 69, 100
capitalism: biodiversity's relation to, 157–60, 189nn6–11; indigenous peoples linked to, 15, 161, 175; naturalization of, 165
Carder, Angela, 87
Cardigan Bay company (Wales), 26
Carsten, Janet, 57, 176
Cavalli-Sforza, Luigi Luca, 162, 164
Cavanah, S., 25
cesarean sections, 86–87
childbirth: male attendee at, 143, 188n17; male initiation linked to, 102, 109; umbilical cord at, 69
child-endangerment laws (California), 80–81, 184–85n2
childgrowth payments, 66–67, 68, 184n22
children: childhood experiences of, 34–35; crops as, 33–34; mortuary feasts for, 144–45; one-blood groupings of, 61–62, 183–84n15; protected by matrilateral payments, 65, 68, 184n22; same-sex parental ties of, 72–73; undifferentiated gender of, 66–67; without parental bond of substance, 78, 95–96, 115–16, 172. *See also* siblingship
Chirac, Jacques, 181n2
CI. *See* Conservation International
Cirad (Montpellier University), 21
Cleon d'Andran (France), 20
cloning: as de-differentiation process, 146–47; of Dolly, 117–18, 186nn1–3; juxtapositional approach to, 14; President's bioethics council on, 120–23, 187n4; therapeutic versus reproductive, 118–20
Columbia Presbyterian Medical Center, 1–3, 179n2
Complexities: Beyond Nature and Nurture (McKinnon and Silverman, ed.), 182–83n2
conception: bridewealth's role in, 64–65, 69–71, 169–70, 172–73, 184nn17,18; death's relation to, 137–38, 148; Kamea understanding of act, 60–61, 172; taboos against, 101–2. *See also* reproduction

Confederation Paysanne (CP), 21
conservation. *See* biodiversity conservation
Conservation International (CI): bioprospecting campaigns of, 159, 189n9; debt-for-nature swaps by, 158–59, 189nn7,8; on indigenous peoples, 160–61, 164, 189n12; mission of, 156–58, 165, 189nn5,6; Papua New Guinea project of, 154, 155–56, 175, 188nn3,4
consumption: biodiversity rhetoric on, 167; cross-cousins linked to, 142–43, 146; in female maturation process, 68–71, 92; Fortes on, 91; at Kamea mortuary feasts, 141–46; as kin-making process, 9–10, 57; of *marita* fruit, 108–9; mother-son habits of, 93–94, 96, 100–101, 185nn7,8,10; sexual connection to, 145, 188n18; at Tangan mortuary feasts, 125–26. *See also* food taboos
containment: bridewealth linked to, 69–71, 72; of bullroarers, 101, 186nn15–17; cross-cousins' role in, 67, 135–37; as mother-child relatedness, 58, 61–62, 95–96, 115–16, 183–84n15, 185n10; of mothers, by offspring, 145, 188n18; of mothers of male initiates, 99–100, 101–2; and recontainment, of siblings, 138–39, 187n14; of siblingship's one-bloodedness, 61, 62 fig.; son's detachment from, 61–62, 90–91, 96, 111, 114–15
Cooley, Judge, 84
cordyline (palm lily), 31
corn, genetically modified, 26
corpses. *See* dead persons
cot'wa (staffs), 105
cousins. *See* cross-cousins
CP (Confederation Paysanne), 21
A Critique of the Study of Kinship (Schneider), 9
crops: as bridewealth, 67; as children, 33–34; as gendered, 33, 182n12; genetically modified, 21–23, 24, 25–26, 181nn3,4,6; intentional eating of, 57; in Kamea gardens, 32; mutagenesis of, 27. *See also* gardens

cross-cousins: consumption linked to, 142–44, 145–46; as marriage negotiators, 67, 135–36, 184n20; at mortuary feasts, 140, 141–42; nose symbol of, 75, 128, 187n13; social differentiation role of, 73, 74–75, 79, 135–36, 138–39, 148, 184n24; taboos between, 75–76; undertaker role of, 131–32, 136–37, 138–39, 187nn8,9, 187nn13,14

cult houses: *marita* ceremony at, 96–97, 108–9, 185–86n12; of women, 99–100, 101–2

culture/nature dialectic. *See* nature/culture dialectic

Current Anthropology (periodical), 48

Cussins, Charis, 76–77, 183n9

Damon, F. H., 64

dance rituals (*marita* ceremony), 104–5, 106 fig., 109–10

Daribi, 60, 65

Darwin, Charles, 5, 44, 45

dead persons: cadaverous fluids of, 137–38, 187n13; cross-cousin's obligation to, 75, 128, 131; Euro-American protections for, 86, 185n3; ornamental memorials to, 129, 130 fig., 140, 188n15; postcolonial rites for, 128–31, 187n6; reproduction by, 50–52, 183nn5,6; smoking rites for, 131–34, 136, 187nn7–9, 187nn11–13

death rituals. *See* mortuary feasts; mortuary rites

debt-for-nature swaps, 158–59, 189nn7,8

decorations. *See* ornaments

Del-Zio, Doris, 1–3, 170, 179n2

Del-Zio, John, 1, 2, 170

descent principle. *See* genealogical model

Descola, Philippe, 28–29, 181–82n8, 182n9

dietary taboos. *See* food taboos

digging stick (*uka*), 35

Dolly (cloned lamb), 117–18, 146–47, 186nn1–3

Douglas, Mary, 27, 94

drug use, as fetal abuse, 80–83, 184–85n2

Duning, Alan, 161

earth oven (*mumu*), 129, 142, 146

eating. *See* consumption

Edwards, Jeanette, 53

Elias, Harry, 81, 184–85n2

embryonic stem cell research. *See* cloning

embryonic totipotency, 146–47

environmentalism. *See* biodiversity conservation

Escobar, Arturo, 160

Ethics Advisory Board (United States), 5

Etoro, 60, 180n6, 183–84n15

Euro-Americans: biodiversity rhetoric of, 157–61, 189nn9–12; biological paradigm of, 5, 176–77; descent principle of, 44; ethical challenges for, 167–68; fetal rights double standard of, 84, 86–88; GMO controversy among, 12–13, 20–21, 23–24, 181n4; informed consent doctrine of, 84–86; kinship formulations of, 7–8, 10, 50, 51–56, 183n10; nature/culture dialectic of, 164–65, 166–67, 176; personhood notion of, 13–14, 90, 112–13, 180n8; trans-species concerns of, 24–27, 44, 181nn5,6; use of term, 16

exchange: at Kamea mortuary feasts, 139–44, 145–46; marriage system of, 67–68, 184nn21,22; in Melanesian mortuary rites, 125–27, 143; of sisters, in marriage, 71–72, 184n23

Fassano, Donna, 49

fatherhood. *See* parents

FDA (Food and Drug Administration), 25

feeding. *See* consumption

females. *See* Kamea women

feminists: on "fetal rights" issue, 83, 84, 86, 87–88; on mutable personhood model, 113–14

fetal abuse cases, 80–83, 184–85n2

fetal rights discourse: criticism of, 84, 86–88, 113; on fetal abuse cases, 80–82, 184–85n2; personhood model of, 88–90, 112–13; on pregnant women's status, 82–83

fieldwork location, 16–17, 19

Fitzgerald, Susan, 24

Flaherty, John, 85
FlavrSavr tomatoes, 21–22
food. *See* consumption; crops
Food and Drug Administration (FDA), 25
food taboos: in anthropology discipline, 91; before mortuary feast, 129, 142, 146; motherless boy's release from, 115; as mother/son unity marker, 92, 94, 100–101, 185nn4,8; of pregnant women, 96; as separation marker, 69, 92–93, 185nn5–7
Ford Motor Company, 157
Fortes, Myer, 91
Foster, Robert, 125–26, 143
Foundation for People and Community Development (Papua New Guinea), 155
Fourth World peoples. *See* indigenous peoples
France, GMO protests in, 20–21
Franklin, Sarah, 147, 166, 171
frozen embryos' rights, 114
funeral practices. *See* mortuary feasts; mortuary rites

Gallagher, Janet, 88, 90
game (*kapul*): as bridewealth, 69–70; hunting of, 32–33, 98; taboos against eating, 94, 100
Garber, Howard, 52
Garber, Jean, 52, 183n7
Garber, Julie, 50, 52, 183n7
gardens: bridewealth from, 67; of initiated males, 107; land-tenure system of, 32, 182n11; location of, 30–31; marital union linked to, 33–34, 36, 182n12. *See also* crops; land
Garner, Dwight, 48
Gawan peoples, 71, 140
gender: at birth, 66; bridewealth's differentiation of, 66–67, 68, 69–71, 72; of crops, 33, 182n12; intentional activation of, 63–64, 172–73; low-tech selection of, 1; male initiation linked to, 94–95, 98–99, 107, 185n8; in mortuary symbolism, 149; of parent-child identification, 72–73; productive domains and, 69

The Gender Gift (Strathern), 183–84n15
"The Genealogical Method of Anthropological Inquiry" (Rivers), 7
genealogical model: biological/social implications of, 44; biotechnology's impact on, 45, 146–47, 170–71; directionality of, 55, 147, 183n9; in parent-child gamete donation, 53–54; of Rivers, 7, 179–80n5; Schneider's critique of, 7–9; siblingship's relation to, 59–60, 183n13
genes: in Christian context, 179n4; connectedness notion of, 51–52, 53–54; evolutionary/lateral transfer of, 45; of indigenous peoples, as other, 162–65, 175; technological transfer of, 21
genetically modified organisms (GMOs): in developing world, 166, 175; examples/uses of, 21–22, 181n3; public protests against, 12–13, 20–21, 22–23, 180–81n1, 181nn2,4; "unintended consequence" notions of, 23–27, 181nn5,6
genetic pollution, 24–27, 181nn5,6
"A Genetic Survey of Vanishing People" (Roberts), 150
George Washington University Hospital, 87
gestational surrogacy, 76–77
Gilbert Islands, 64
Ginsburg, F., 180n7
GMOs. *See* genetically modified organisms
Godelier, Maurice, 16, 18, 184n23
golden rice, 22
Good Samaritan law, 85
"good woman" (*awati apaka*), 33
Gothard, Samantha, 54
grass skirts (*pulpuls*), 106
Great Britain. *See* United Kingdom

Hagahai cell lines, 154, 175
hair rituals, 129, 142
haka (bamboo), 69
Harris, Paul, 52
Harris poll (1969), 3–4
Hawabango (Catholic mission station), 103
Hayden, Cori, 163–64, 181n5
headdresses, 104, 112 fig.

National Institute of Health (NIH), 154
nature/culture dialectic: in biodiversity
    context, 153–54, 164–65, 188n1;
    biotechnology's impact on, 5, 152–53,
    166–67, 175–77; Descola on, 28–29,
    181–82n8, 182n9; of gender, 149; glob-
    alizing shift in, 151–52; Kamea per-
    spective on, 6, 29–30; of North Amer-
    ican kinship, 7–9; of relatedness
    kinship model, 9–10
Nedelsky, Jennifer, 114
Nelkin, D., 179n4
Neo-Melanesian (*tok pisin*) language, 17,
    18, 180n11
Nerac (France), 21
New Caledonians, 124
New England Journal of Medicine survey, 86
New Guinea. *See* Papua New Guinea
new reproductive technologies (NRTs). *See*
    reproductive technologies
New York Times, 25
NIH (National Institute of Health), 154
nka apaka ("my woman"), 74
nka dato ("my brother"), 74
nka nabi ("my sister"), 74
nka oka ("my man"), 74
North Americans. *See* Euro-Americans
nose *(hiyma),* 75, 128, 187n13
nose piercing, 35, 96, 97, 98, 100–101, 102
Nova (British magazine), 4
Novataris AG (Swiss firm), 21
NRTs (new reproductive technologies). *See*
    reproductive technologies
nuclear transfer technique (SCNT), 118,
    146–47, 186n2
nuwa (shell ornaments), 106–7

Oates, L., 74
Oates, W., 74
oka. See Kamea men
one-bloodedness *(hinya avaka):* cross-
    cousins' relationship to, 136–37, 138–
    39, 187n14; Kamea notion of, 13; of
    siblingship, 61, 62 fig., 95; ungen-
    dered connotation of, 66–67, 71
Operation Cremation Monsanto (India),
    23

Orientalism (Said), 16
The Origin of Species (Darwin), 5, 44, 45
origins myth, 36–39, 109, 137–38
ornaments: *he'aka,* for mourners, 129, 130
    fig., 140, 188n15; at *marita* initiation,
    104, 106–7, 112 fig.

pa'a (type of sorcery), 71
palm lily (cordyline), 31
pandanus fruit *(marita),* 96, 100, 108–9,
    129, 142, 146
pandanus nut trees *(anaka),* 33–34
pannga (witches), 130
Papua New Guinea: Conservation Interna-
    tional's project in, 154, 155–57, 158,
    175, 188nn3,4; fieldwork location in,
    16–17, 19; Kamea habitat in, 30, 31
    fig.; map of Kamea region, 17 fig.;
    map of Lakekamu Basin, 156 fig. *See
    also* highland peoples
parent-child gamete donation, 52–55
parents: Britain's definition of, 47; in child-
    parent gamete donation, 55; gesta-
    tional surrogates as, 76–77; in mis-
    taken IVF cases, 46–49; in parent-
    child gamete donation, 52–54;
    posthumous reproduction by, 50–52,
    183nn5,6; same-sex ties to, 72–73;
    without substance bond to children,
    10, 61–62, 78, 95–96, 115–16, 172,
    180n6, 183–84n15
Parkin, Robert, 165
Parry, J., 125, 149
personhood: biological reproduction of,
    13–14, 51–52, 180n8, 183n7; in
    Canaque naming practices, 124; in
    cloning debate, 120–22, 123, 187n4;
    common law principles of, 84–86,
    185n3; denied to pregnant women, 84,
    86–88; Euro-American model of, 112–
    13; of fetus, 89–90; of frozen embryos,
    114; as mutable/connective, 113–14
pest control, 21–22, 181n3
Pfannenstiel, Diane, 81
pigs. *See* pork
pipia (rubbish) piles, 182n12
pitpit (cane), 32, 69, 100

Text: 11.25/13.5 Adobe Garamond
Display: Adobe Garamond
Indexer: Patricia Deminna
Illustrator: Bill Nelson
Compositor: Binghamton Valley Composition, LLC
Printer and binder: Maple-Vail Manufacturing Group